Biosensors:
an Introduction

Dr B. R. Eggins is lecturer in physical and analytical chemistry at the University of Ulster. His research interests include electrochemistry and photoelectrochemistry as well as biosensors. He has lectured in Canada, USA and Europe. As well as over 50 research papers he has published *Chemical Structure and Reactivity*. He is a Fellow of the Royal Society of Chemistry.

Biosensors: an Introduction

Brian R. Eggins
University of Ulster at Jordanstown

⊛WILEY 田TEUBNER

A Partnership between John Wiley & Sons and B. G. Teubner Publishers

Chichester · New York · Brisbane · Toronto · Singapore · Stuttgart · Leipzig

Copyright © 1996 by John Wiley & Sons Ltd and B.G. Teubner

John Wiley & Sons Ltd	B.G. Teubner
Baffins Lane	Industriestraße 15
Chichester	70565 Stuttgart (Vaihingen)
West Sussex PO19 1UD	Postfach 80 10 69
England	70510 Stuttgart
	Germany

National Chichester (01243) 779777   *National* Stuttgart (0711) 789 010
International +44 1243 779777   *International* +49 711 789 010

Reprinted March 1997

Designations used by companies to distinguish their products are often claimed
as trademarks. In all instances where John Wiley & Sons Ltd and B.G. Teubner
are aware of a claim, the product names appear in initial capital or all capital letters.
Readers, however, should contact the appropriate companies for more complete
information regarding trademarks and registration.

Other Wiley Editorial Offices

John Wiley & Sons, Inc., 605 Third Avenue,
New York, NY 10158-0012, USA

Brisbane • Toronto • Singapore

Other Teubner Editorial Offices

B.G. Teubner, Verlagsgesellschaft mbH, Johannisgasse 16,
D-04103 Leipzig, Germany

Library of Congress Cataloging-in-Publication Data

Eggins, Brian R.
   An introduction to biosensors / Brian R. Eggins.
      p.   cm.
   Includes bibliographical references and index.
   ISBN 0-471-96285-6 (alk. paper)
   1. Biosensors.   I. Title.
   R857.B54E35   1996
   681'.2—dc20   95-40660
                                                              CIP

Die Deutsche Bibliothek—CIP-Einheitsaufnahme

Eggins, Brian R.:
Biosensors : an introduction / Brian R. Eggins. – Stuttgart ;
Leipzig ; Teubner ; Chichester ; New York ; Brisbane ; Toronto
; Singapore : Wiley, 1996
   ISBN 3-519-02117-X (Teubner)
   ISBN (falsch) 0-471-96285-6 (Wiley)

WG: 32;30	DBN 94.683346.X	96.02.13
7464	nh	

British Library Cataloguing in Publication Data

A catalogue record for this book is available from the British Library

ISBN Wiley 0 471 96285 6
ISBN Teubner 3-519-02117-X

Typeset in 10/12pt Times by Mackreth Media Services, Hemel Hempstead, Herts
Printed and bound in Great Britain by Biddles Ltd, Guildford, Surrey
This book is printed on acid-free paper responsibly manufactured from sustainable forestation,
for which at least two trees are planted for each one used for paper production.

To Chrissie

Contents

Preface

The need and the idea for this book came out of the author's involvement in delivering a lecture course on biosensors to final year students of the highly successful Honours Degree in Applied Biochemical Sciences at the University of Ulster at Jordanstown. This four-year course was designed to meet the needs of local industry in Northern Ireland. It includes courses in chemistry, biochemistry, analytical methods, industrial studies and computing, business studies, projects in science and business studies, with the third year spent on placement in industry, usually in the UK or Ireland.

The 12 lecture + 12 hour laboratory course in biosensors is part of a module called 'special topics', which also includes bioinorganic chemistry and chemotherapeutics.

There were not many books covering biosensors, and those that did exist were mostly specialised, but good reference books for further reading.

A reasonable background knowledge of electroanalytical methods is needed as a prerequisite for a study of biosensors. In our course this material is covered as part of the advanced analytical methods module. A substantial part of this book is devoted to electrochemical transducers.

The book commences with an introductory chapter about the nature of biosensors, their components and applications. This is followed by a chapter surveying the groups of biological components involved in biosensors. Chapter 3 covers the various ways in which the biological components can be attached to the transducers, i.e. immobilised.

Following that are three chapters on different types of transducers, with examples of how they are used in biosensors, Chapter 4 covers electroanalytical methods, which are the ones having the most use in biosensors so far. The photometric methods described in Chapter 5 have so far been rather expensive for use in sensors, but are developing faster and have a promising future. Chapter 6 covers miscellaneous methods such as piezoelectric and thermistor-based transducers.

It may be easy to develop novel biosensors, but it is not so easy to check out their performance criteria. Chapter 7 describes the principal performance characteristics of various biosensors and how they are likely to vary with different biological components, different transducers and different modes of immobilisation. These principles are continued throughout the rest of the book.

Chapter 8 illustrates the application of biosensors to a variety of important analytical situations at the laboratory level. In Chapter 9 are set out the practical experiments used by our students, together with some sample results. The experiments illustrate the use of glucose oxidase with a ferrocene mediator in a voltammetric mode, a potentiometic urea sensor based on a pH electrode and the differential-pulse voltammetric determination of catechols in beers using a banana-based electrode.

The final chapter, Chapter 10, discusses the problems of the commercial production of biosensors and various areas of application such as in medicine, biotechnology, the environment and defence.

Each chapter has a number of references, many of which are not specifically mentioned in the text and are for further reading, including books, review articles and original papers. The references are not intended to be comprehensive. Some important general books and some review articles are listed at the end of this preface.

I would like to thank a number of people for their help with this book. First, Rosemary Clelland of the Graphics Department, University of Ulster, for her considerable graphical work with the figures and diagrams. Then I thank several generations of students who have listened to my lecture course, and patiently tested different versions of the experiments described in Chapter 9, finally producing the results presented. In particular, I thank the students who carried out their projects on topics involving biosensors. I thank my colleagues for their help and encouragement. I thank Medisense Ltd, Abingdon, for their kind gift of an Exactech® glucose biosensor. Lastly, I thank Professor Allen Hill of Oxford University for the lecture he gave on biosensors at the University of Ulster in 1992, which inspired the idea of writing this book and for his helpful advice at a later stage.

GENERAL REFERENCES

A. E. G. Cass (Ed.) (1990) *Biosensors: A Practical Approach*, Oxford University Press, Oxford.
G. G. Guilbault (1984) *Analytical Uses of Immobilised Enzymes*, Marcel Dekker, New York.
G. G. Guilbault and M. Mascini (Eds) (1988) *Analytical Uses of Immobilised Biological Compounds for Detection, Medical and Industrial Uses*, NATO ASI Series, Reidel, Dordrecht.

E. A. H. Hall (1990) *Biosensors*, Open University Press, Milton Keynes.
A. P. F. Turner, I. Karube and G. S. Wilson (Eds) (1987) *Biosensors: Fundamentals and Applications*, Oxford University Press, Oxford.

ARTICLES

M. C. Goldschmidt (1993) 'Biosensors: blessing or bottleneck?', *Rapid Methods Autom. Anal.*, **2**, 9.
D. Griffiths and G. Hall (1993) 'Biosensors — what real progress is being made?', *TIBTECH*, **11**, 122.
J. D. Newman and A. P. F. Turner (1994) 'Biosensors: The analyst's dream?', *Chem. Ind.*, 16th May, 374.
M. Romito (1993) 'Biosensors: diagnostic workhorses of the future?' *S. Afr. J. Sci.*, **89**, 93.
F. W. Scheller, F. Schubert, R. Renneberg, H.-G. Muller, M. Janchen and H. Weise (1985) 'Biosensors: trends and commercialisation', *Biosensors*, **1**, 135.
A. P. F. Turner (1994) 'Biosensors', *Current Opinion Biotechnol.*, **5**, 49.
P. Vadgama and P. W. Crump (1992) 'Biosensors: Recent trends', *Analyst*, **117**, 1657.

Chapter 1

Introduction

1.1 WHAT IS A BIOSENSOR?

Well, we all have two of them—our noses and our tongues! But first let us ask a more general question—what is a sensor?

One of the best known types of sensor is the litmus paper test for acids and alkalis, which gives a qualitative indication by means of a colour reaction, of the presence or absence of acid. A more precise method of indicating the degree of acidity is the measurement of pH either by a more extended use of colour reactions in special indicator solutions or even pH papers. However, the best method of measuring acidity is the use of the pH meter, which is an electrochemical device giving an electrical response which can be read by a needle moving on a scale or on a digital readout device or by input to a microprocessor.

In these methods, the *sensor* which responds to the degree of acidity is either a chemical—the dye litmus, or a more complex mixture of chemical dyes in pH indicator solutions—or the glass membrane electrode in the pH meter.

The chemical or electrical response then has to be converted into a signal which we can observe, usually with our eyes. With litmus this is easy. A colour change is observed, because of the change in the absorbance of visible light by the chemical itself, which is immediately detected by our eyes in a lightened room. In the case of the pH meter, the electrical response (a voltage change) has to be converted i.e. *transduced* = led through, into an observable response—movement of a meter needle or a digital display. The part of the device which does this conversion is called a *transducer*.

In a *biosensor*, the sensing element which responds to the substance being measured is biological in nature. It has to be connected to a transducer of some sort so that a visually observable response occurs. Biosensors are generally concerned with sensing and measuring particular chemicals which need not be biological components themselves, although sometimes they are. We shall usually refer to them as the *substrate*, although the more general term *analyte* is often used. Figure 1.1 shows schematically the general arrangement of a biosensor.

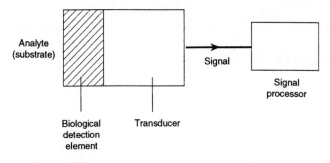

Figure 1.1   Schematic layout of a biosensor

What we detect with the nose—the smell—is small quantities of chemicals. The nose is an exceptionally sensitive and selective instrument which is very difficult to emulate artificially. It can distinguish between many different chemical substances qualitatively and can give a general idea of quantity down to very low limits. The chemicals to be detected pass through the olfactory membrane to the olfactory bulbs, which are the biological components that sense the substrate. The response is an electrical signal that is transmitted to the brain via the olfactory nerves. The brain transduces this response into the sensation we know as smell. The tongue operates in a similar way.

Figure 1.2 shows a schematic diagram of the nasal olfactory system showing the comparison with our generalised biosensor. The nostrils collect the smell sample, which is then sensed by the olfactory membrane, which is the equivalent of the biological component. The responses of the olfactory membrane are then converted by the olfactory nerve cell, which is equivalent to the transducer, into electrical signals which pass along the nerve fibre to the brain for interpretation. Thus the brain acts as a microprocessor, turning the signal into a sensation which we call smell.

A *Biosensor* can be defined as *a device incorporating a biological sensing element connected to a transducer*.

A *Transducer* converts an observed change (physical or chemical) into a measurable signal, usually an electronic signal whose magnitude is proportional to the concentration of a specific chemical or set of chemicals. It

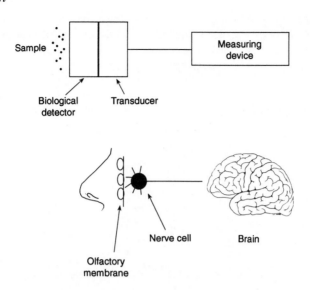

Figure 1.2 The nose as a biosensor

is an apparently alien marriage of two contrasting disciplines. It combines the specificity and sensitivity of biological systems with the computing power of the microprocessor.

1.2 THE FIRST BIOSENSORS

These were often called *enzyme electrodes* and were first described by Clark and Lyons (1962) for the determination of glucose.

This is by far the most studied and developed biosensor application. Glucose is of special importance because of its involvement in human metabolic processes. In particular sufferers from diabetes mellitus do not produce sufficient insulin in their pancreases to control adequately the level of glucose in their blood. Doses of insulin have to be administered and it is vital that the diabetic regularly monitors the level of glucose in the blood. Previously, substantial blood samples had to be taken and analysed in a medical laboratory with consequent time delays and consequent uncertainty about the diabetic's condition.

With the currently available glucose biosensors (such as the ExacTech, produced by Medisense), the patient himself or herself can extract one small drop of blood and obtain a direct digital readout of the glucose concentration inside 1 min.

Glucose biosensors are based on the fact that the enzyme glucose oxidase catalyses the oxidation of glucose to gluconic acid. In the early biosensors oxygen was used as the oxidising agent. The consumption of oxygen was followed by electrochemical reduction at a platinum electrode, as in the Clark oxygen electrode (invented in 1953), shown in Figure 1.3.

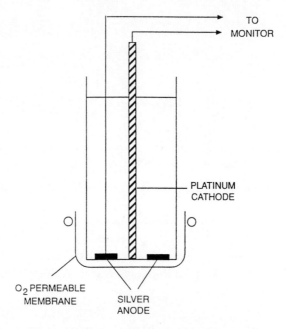

Figure 1.3   A Clark oxygen electrode. Reproduced by permission of the Open University Press from Hall (1990)

The glucose oxidation reaction, catalysed by glucose oxidase (GOD) is

$$\text{glucose} + O_2 + H_2O \xrightarrow{\text{GOD}} \text{gluconic acid} + H_2O_2$$

At the electrode:

$$O_2 + 2e^- + 2H^+ = H_2O_2$$

A voltage of -0.7 V is applied between the platinum cathode and the silver anode, sufficient to reduce the oxygen, and the cell current, which is proportional to the oxygen concentration, is measured. The concentration of glucose is then proportional to the decrease in current (oxygen concentration). The oxygen electrode has an oxygen-permeable membrane (such as PTFE, polythene or Cellophane) covering the electrodes. A layer of enzyme (glucose oxidase) is placed over this and held in place with a second membrane such as cellulose acetate, as shown in Figure 1.4.

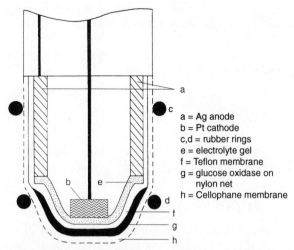

Figure 1.4   A glucose biosensor based on the Clark oxygen electrode. Reproduced by permission of the Open University Press from Hall (1990)

A schematic diagram showing the general layout (which is applicable to many biosensors) is shown in Figure 1.5.

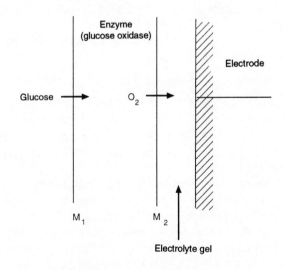

Figure 1.5   Schematic layout of a Clark biosensor for glucose

The substrate (glucose solution) and oxygen can penetrate the first membrane to react with the enzyme to form products. Only the oxygen remaining can penetrate the second membrane to be measured at the electrode. This biosensor was first patented in 1970 and marketed by the Yellow Springs Instrument Company in 1974.

1.3 UREA BIOSENSORS

Another early biosensor was that to determine urea (the major component of urine). Urea is hydrolysed by the enzyme urease to form ammonia and carbon dioxide:

$$CO(NH_2)_2 + H_2O \xrightarrow{\text{urease}} CO_2 + 2NH_3$$

The concentration of ammonia is monitored by means of an ammonium ion-selective electrode, with which the relative voltage of the electrode is measured at close to zero current. This voltage is proportional to the logarithm of the concentration of ammonia, which is directly proportional to the concentration of urea.

The ammonium ion-selective electrode is really a modified glass pH electrode. The enzyme is mixed with a gel and coated on to a nylon net membrane covering the electrode. It is held in place with a second (dialysis) membrane. This biosensor was originally developed by Guilbault and Montalvo (1969).

A simple urea electrode can be made in any laboratory using an ordinary glass pH electrode (see Chapter 9).

1.4 THE BANANATRODE

A very simple biosensor can be made using a banana. It was described for the determination of dopamine, an important brain chemical (Sidwell and Rechnitz, 1985; Wang and Lin, 1988).

Many experiments have been conducted by implanting electrodes in live animal brains to monitor the change in dopamine levels with various activities. These experiments would be greatly enhanced if the dopamine biosensor was used, as it is much more specific to dopamine, avoiding the interference from ascorbic acid.

One simply mixes a small amount of banana with graphite powder and liquid paraffin and places some of the mixture in an electrode cup to make the biosensor (see Chapter 9).

Dopamine is a catechol derivative. The enzyme polyphenol oxidase in banana catalyses the oxidation of the dihydroxy form of dopamine to the quinone form using ambient oxygen. The electrochemical reduction of the quinone back to the dihydroxy form causes a current to flow, which is directly proportional to the concentration of dopamine. The biosensor works equally well with catechol (1,2-dihydroxybenzene). Figure 1.6 shows the reaction.

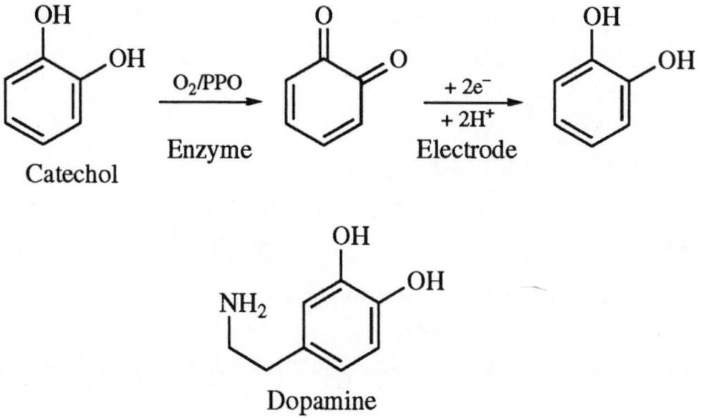

Figure 1.6   Reaction of catechol in a biosensor

1.5   BIOSENSORS MAY BE CONSIDERED UNDER A NUMBER OF HEADINGS

1.5.1   The Analyte or Substrate

There are now so many of these that it is impossible to generalise about categories of substrate. Virtually any substance that is consumed or produced in a biochemical process can in principle be analysed by a biosensor if one can be constructed. A few diverse examples can be given:

sugars	cholesterol
urea	penicillins
creatine	paracetamol
ethanol	aspirin
glutamic acid	TNT
lactic acid	many amino acids
phosphate	

1.5.2   The Biological Component

The importance of the biological component is that its interaction with the substrate is highly specific to that substrate alone, thus avoiding interferences from other substances which plague many analytical methods. It may catalyse a reaction involving the substrate (enzyme) or it may bind selectively to the substrate. The most common component is the *enzyme*, although other components containing enzymes are often very suitable. These include *microorganisms* such as yeasts and bacteria and *tissue material* such as the banana, already mentioned, and liver.

Antibodies used in various modes can be used to bind substrates selectively. *Nucleic acids* sometimes have applications. Details are given in Chapter 2.

1.5.3   Methods of Immobilisation

The biological component has to be intimately connected to the transducer and a number of methods of doing this have evolved.

(i)   The simplest is *adsorption* on to a surface.

(ii)   *Microencapsulation* is the word used for trapping between membranes—one of the earliest methods as illustrated with the glucose and urea biosensors.

(iii) *Entrapment.* The biological component is trapped in a matrix of a gel or a paste or a polymer (as with the banana). This is a very popular method.

(iv)   *Covalent attachment.* Covalent chemical bonds are formed between the biological component and the transducer.

(v)   *Cross-linking.*
       A bifunctional agent is used to bond chemically the transducer and the biological material. This is often used in conjunction with other methods such as (i) or (iii).

More details about these procedures are discussed in Chapter 3.

1.5.4   Transducers—The Detector Device

Most biosensors have been constructed around electrochemical transducers, but a number of other types have been used, and photometric transducers in particular are growing in importance. However, as long as our microprocessors are driven by electrons, the directness of an electrical response will tend to have maximum appeal. One day photon-driven devices will make most of our electrical appliances obsolete—starting with telephones?

Transducers can be subdivided into various types, as follows.

(i)   Electrochemical transducers
       (a) Potentiometric. These involve the measurement of the e.m.f. (potential) of a cell at zero current. The e.m.f. is proportional to the logarithm of the concentration of the substance being determined.
       (b) Voltammetric. An increasing (decreasing) potential is applied to the cell until oxidation (reduction) of the substance to be analysed

occurs and there is a sharp rise (decrease) in the cell current to give a peak current. The height of this peak current is directly proportional to the concentration of the electroactive material.

If the appropriate oxidation (reduction) potential is known, one may step the potential directly to that value and observe the current. This mode is known as *amperometric*.

(c) *Conductimetric.* Solutions containing ions conduct electricity. The magnitude of this conductance may change due to a chemical or a biochemical reaction. The relationship between conductance and concentration depends on the nature of the reaction. Measurement can be very simple.

(ii) *FET-based sensors.* Miniaturisation can sometimes be achieved by constructing one of the above types on a field effect transistor (FET) chip. This has mainly been used with potentiometric sensors, but could also be used with voltammetric or conductimetric sensors.

(iii) *Optical transducers.* These have taken on a new lease of life with the development of fibre optics, allowing greater flexibility and miniaturisation. Techniques used include absorption spectroscopy, fluorescence spectroscopy, luminescence spectroscopy, internal reflection spectroscopy, surface plasmon resonance and light scattering.

(iv) *Piezoelectric devices.* These devices involve the generation of electric currents from a vibrating crystal. The frequency of vibration is affected by the mass of material adsorbed on its surface, which could be related to an active biochemical reaction.

(v) *Surface acoustic waves.*

(vi) *Thermal methods.* All chemical and biochemical processes involve the production or absorption of heat. This heat can be measured by sensitive thermistors and hence be related to the amount of reaction.

Details of these transducers are given in Chapters 4 and 5.

1.5.5 Performance Factors

1.5.5.1 SELECTIVITY

This is the most important characteristic of biosensors—the ability to discriminate between different substrates. This is a function of the biological component, principally, although sometimes the operation of the transducer contributes to selectivity.

1.5.5.2 SENSITIVITY RANGE

This usually needs to be sub-millimolar, but in special cases can go down to the femtomolar (10^{-15} M) range.

1.5.5.3 ACCURACY

This is usually around ±5%.

1.5.5.4 NATURE OF SOLUTION

Conditions such as pH, temperature and ionic strength must be considered.

1.5.5.5 TIMES

(i) The *response time* is usually much longer than with chemical sensors. It may be 30s or longer. (ii) The *recovery time* is the time before a biosensor is ready to analyse the next sample. It must not be too long—not more than a few minutes. (iii) The *working lifetime* is usually determined by the instability of the biological material. It can vary from a few days to a few months. The Exactech glucose biosensor is usable for over 1 year. Further details are given in Chapter 7.

1.6 SCOPE

The possible range of analytes is almost limitless. They range from simple inorganic ions such as nitrate, through simple molecules such as carbon monoxide, methane and methanol, to complex biomolecules such as NAD. They are limited only by the ingenuity of scientists to discover and match appropriate enzymes, bacteria, antibodies, etc. New ones are being discovered regularly. Mutations and genetic engineering open the possibilities of growing new strains or mutants which might have new applications. However, discovering and developing a biosensor in the laboratory is a long way from applying it in a practical situation, let alone developing it for mass manufacture and marketing. The most successful biosensor by far which is marketed by a number of different companies is the glucose biosensor. Others which have been developed close to the marketing level are for cholesterol, aspirin and paracetamol.

1.7 APPLICATIONS

In this section we shall survey a number of areas of potential application of biosensors.

1.7.1 Health Care

As the reader may have gathered, this is the main area of application and potential application. All the examples referred to in Section 1.6 relate to health care. Measurements of blood, gases, ions and metabolites are often necessary to show the patient's metabolic state—especially for patients in hospital, and even more so if they are in intensive care. Many of these substrates have been determined by samples of urine and blood being taken away to a medical analytical laboratory for classical analysis, which may not be complete for hours or even days. The use of on-the-spot biosensors could enable analytical results to be obtained within minutes at most. The ExacTech glucose biosensors gives a reading in 30 s. This would obviate the need for *en suite* analytical units with specialist medical laboratory scientists. A trained nurse would be competent to carry out biosensor tests at the bedside. They could also be used in casualty departments (for example, for checking whether a suspected drug overdose might be due to paracetamol or aspirin) and for general clinical monitoring. They could also be used in GP-manned health centres or even by the patients themselves at home—particularly in the case of diabetics. Table 1.1 shows some common assays needed for diagnostic work with patients. Most of these could potentially be done with biosensors.

Table 1.1   Commonly required instant assays in diagnosis of patients

Ions	Gases	Drugs	Substrates	Enzymes
Sodium	Oxygen	Paracetamol	Glucose	Creatine kinase
Potassium	Ammonia	Salicylate	Cholesterol	Amylase
Calcium	Carbon dioxide		Creatinine	Aspartate
pH				Aminotransferase

A potential dream application is to have an implanted biosensor for continuous monitoring of a metabolite. This might then be linked via a microprocessor to a controlled drug-delivery system through the skin. This would be particularly attractive for chronic conditions such as diabetes. The blood glucose would be monitored continuously and, as the glucose level reached a certain level, insulin would be released into the patient's blood stream automatically. This system is sometimes referred to as an *artificial pancreas*. This would be far more beneficial for the patient than the present injection of larger discrete doses every few hours.

1.7.2   Control of Industrial Processes

Biosensors can be used in various aspects of fermentation processes in three ways: (i) off-line in a laboratory; (ii) off-line, but close to the operation; and

(iii) on-line, in real time. At present, the main real-time monitoring is confined to pH, temperature, CO_2 and O_2. Biosensors which monitor a much wider range of direct reactants and products are available, such as various sugars, yeasts, malt and alcohols and perhaps also undesirable byproducts. Such monitoring could result in improved product quality, increased product yields, checks on tolerance of variations in quality of raw materials, optimised energy efficiency, i.e. improved plant automation, and less reliance on human judgement. There are a wide range of applications in the food and drinks industry generally.

1.7.3   Environmental Monitoring

There is an enormous range of potential analytes in air, water, soils and other situations. Such things as BOD, acidity, pesticides, fertilisers, industrial wastes and domestic wastes require extensive analyses. Continuous real-time monitoring is required for some substances and occasional random monitoring for others. In addition to the obvious pollution applications, farming, gardening, veterinary science and mining are potential areas where biosensors could be used for environmental monitoring.

1.8   REFERENCES

L. C. Clark, Jr (1987) 'The enzyme electrode', in A. P. F. Turner, I. Karube and G. S. Wilson (Eds), *Biosensors: Fundamentals and Applications*, Oxford University Press, Oxford, Chap. 1, pp. 3–12.

L. C. Clark, Jr and C. Lyons (1962) 'Electrode systems for continuous monitoring in cardiovascular surgery', *Ann. N.Y. Acad. Sci.*, **102**, 29.

L. C. Clark, Jr, R. Wolf, D. Granger and Z. Taylor (1953) 'Continuous recording of blood oxygen tensions by polarography', *J. Appl. Physiol.* **6**, 189.

G. G. Guilbault and J. Montalvo (1969) *J. Am. Chem. Soc.*, **91**, 2164.

E. A. H. Hall (1990) *Biosensors*, Open University Press, Milton Keynes.

P. D. Home and K. G. M. M. Alberti (1987) 'Biosensors in medicine: the clinician's requirements', in A. P. F. Turner, I. Karube and G. S. Wilson (Eds), *Biosensors: Fundamentals and Applications*, Oxford University Press, Oxford, Chap. 36, pp. 723–736.

J. McCann (1987) 'Exploiting biosensors', in A. P. F. Turner, I. Karube and G. S. Wilson (Eds), *Biosensors: Fundamentals and Applications*, Oxford University Press, Oxford, Chap. 37, pp. 737–746.

J. S. Sidwell and G. A. Rechnitz (1985) 'Bananatrode—an electrochemical sensor for dopamine', *Biotechnol. Lett.*, **7**, 419.

J. Wang and N. S. Lin (1988) 'Mixed plant tissue–carbon paste bioelectrode', *Anal. Chem.* **60**, 1545.

Chapter 2

Biological Elements

2.1 INTRODUCTION

Biological elements provide the major selective element in biosensors. They must be substances that can attach themselves to one particular substrate but not to others. Four main groups of materials can do this:

(i)   Enzymes.
(ii)   Antibodies.
(iii) Nucleic acids.
(iv)   Receptors

The biological elements most regularly used are *enzymes*. These may be used in a purified form, or may be present in microorganisms or in slices of intact tissue. They are biological catalysts for particular reactions and can bind themselves to the specific substrate. This catalytic action is made use of in the biosensor.

Antibodies have a different mode of action. They will bind specifically with the corresponding antigen, to remove it from the sphere of activity, but they have no catalytic effect. Despite this, they are capable of developing ultra-high sensitivity in biosensors. Considerable ingenuity is often needed to involve them with the substrate and to provide a signal for the transducer to measure.

Nucleic acids have been much less used so far. They operate selectively because of their base-pairing characteristics. They have great potential utility in identifying genetic disorders, particularly in children.

Inside the lipid bilayer plasma membrane surrounding a cell are proteins that traverse the full breadth of the membrane and which have molecular recognition properties. They are known as *receptors*. They are difficult to isolate, but will bind solutes with a degree of affinity and specificity matching antibodies.

2.2 ENZYMES

An enzyme is a large, complex macromolecule, consisting largely of protein, usually containing a *prosthetic* group, which often includes one or more metal atoms. In many enzymes, especially in those used in biosensors, the mode of action involves oxidation or reduction which can be detected electrochemically. We shall not discuss here the detailed mode of action of enzymes, which can be found in any standard biochemistry text, but just remind ourseleves of the basic enzyme catalysis mechanism:

$$S + E \; \underset{k_{-1}}{\overset{k_1}{\rightleftharpoons}} \; ES \; \overset{k_2}{\longrightarrow} \; E + P \tag{2.1}$$

Where S = substrate(s), E = enzyme, ES = enzyme–substrate complex and P = product(s); for example, S = glucose + oxygen, E = glucose oxidase (GOD) and P = gluconic acid + hydrogen peroxide:

$$\text{glucose} + O_2 + \text{GOD} \longrightarrow ES \longrightarrow \text{gluconic acid} + H_2O_2$$

Let us apply the steady-state approximation of kinetic theory to the system shown in equation 2.1. This approximation simply assumes that, during most of the time of reaction, the concentration of the enzyme–substrate complex is steady, i.e. constant, so the rate of formation of the complex from its component is balanced by the rate of its breakdown back to enzyme and forward to products. Thus,

$$\text{rate of formation of complex} = k_1 \, [S][E] - k_{-1}[ES]$$

$$\text{rate of breakdown of complex} = k_2[ES]$$

These rates are equal and opposite because of the steady-state approximation. Therefore,

$$k_1 \, [S][E] - k_{-1} \, [ES] - k_2[ES] = 0$$

We describe the enzyme concentration in terms of the total $[E_0]$ rather than the unknown [E], so that $[E_0] = [E] + [ES]$. Then,

$$k_1[S][E_0] - k_1[S][ES] - k_{-1}[ES] - k_2[ES] = 0$$

If we solve this equation for [ES], we obtain

$$[ES] = \frac{k_1[E_0][S]}{k_{-1} + k_2 + k_1[S]}$$

If we now put $K_M = (k_{-1} + k_2)/k_1$, where K_M is the *Michaelis constant*, we obtain

$$[ES] = \frac{[E_0][S]}{K_M + [S]}$$

Then the overall rate of reaction (rate of formation of products) is given by the *Michaelis–Menton equation*:

$$v = \frac{d[P]}{dt} = -\frac{d[S]}{dt} = k_2[ES] = \frac{k_2[E_0][S]}{K_M + [S]}$$

When $[S] \gg K_M$, a maximum value of the rate constant, V_{max}, is reached, so that $V_{max} = k_2[E_0]$, and when $[S] \ll K_M$, $v = V_{max}/2$. This is shown in Figure 2.1, which is a curve. It is experimentally more convenient to plot the data in straight-line form. This can be donw by inverting the Michaelis–Menton equation:

$$1/v = \frac{K_M + [S]}{k_2[E_0][S]} = \frac{K_M}{k_2[E_0][S]} + \frac{1}{V_{max}}$$

This is the *Lineweaver–Burk* plot. When $1/v$ is plotted against $1/[S]$, a straight line is obtained with a slope K_M/V_{max} and intercept $1/V_{max}$, hence both K_M and V_{max} can be obtained. In practice, many more steps may be involved and the influence of inhibitors would need to be considered.

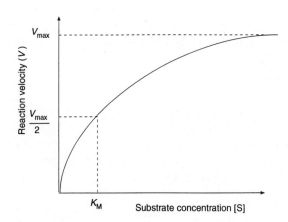

Figure 2.1 Dependence of reaction rate on substrate concentration for an enzyme-catalysed reaction at constant enzyme concentration.

2.3 EXAMPLES OF ENZYME BIOSENSORS

2.3.1 Urea

The hydrolytic breakdown of urea is catalysed by the enzyme *urease* to give ammonia and carbon dioxide:

$$(NH_2)_2CO + H_2O \xrightarrow{\text{urease}} 2NH_3 + CO_2$$

The reaction is usually followed potentiometrically with an ammonia ion-selective electrode (ISE), which can detect ca 10^{-6} M ammonia. The performance criteria for this biosensor are as follows:

Range: $3 \times 10^{-5} - 5 \times 10^{-2}$ M.
Response time: 1–5 min.
Recovery time: 5–10 min (this can be improved by better immobilisation to permit up to 20 assays per hour).
Lifetime: 60 days.

There are alternative modes of operation of this electrode which can use an ammonium ISE or a carbon dioxide electrode or even a pH electrode. Of course, the operational pH needs to be controlled very carefully with appropriate buffers to optimise operation with a particular electrode.

2.3.2 Glucose

This is the most extensively studied biosensor and the most commercially developed. The original form involved the oxidation of glucose by molecular oxygen, catalysed by the enzyme glucose oxidase (GOD) to give gluconic acid and hydrogen peroxide;

$$\text{glucose} + O_2 \xrightarrow{\text{GOD}} \text{gluconic acid} + H_2O_2$$

The reaction was originally followed with a Clark oxygen electrode, which monitors the decrease in oxygen concentration amperometrically. The performance criteria are as follows:

Range: <1–30 mM.
Response time: 1–1.5 min.
Recovery time: 30 s.
Lifetime: several months.

2.3.3 Advantages and Disadvantages of Enzymes

Advantages
  (i) They bind to the substrate.
 (ii) They are highly selective.
(iii) They have catalytic activity, thus improving sensitivity.
 (iv) They are fairly fast acting.
  (v) They are the most commonly used biological component.

Disadvantages
  (i) They are expensive. The cost of extracting, isolating and purifying enzymes is very high, and sometimes the cost of the source of the enzyme may be high. However, a very wide range of enzymes are available commercially, usually with well defined and assayed characteristics.
 (ii) There is often a loss of activity when they are immoblised on a transducer.
(iii) They tend to lose activity, owing to deactivation, after a relatively short period of time.

2.4 TISSUE MATERIALS

Plant and animal tissues may be used directly with minimal preparation. Generally tissues contain a multiplicity of enzymes and thus may not be as selective as purified enzymes. However, the enzymes exist in their natural environment and so are less subjective to degradation. Hence the sensors are likely to have a longer lifetime. On the other hand, the response may be slower as there is more tissue material for the substrate to diffuse through. This material may also dilute the effect of the enzymes. We shall look at a number of examples, comparing the performance criteria of biosensors made from tissue material with those made from pure enzymes.

Both microorganisms and tissues are enzyme-containing materials, but the different environments surrounding the enzymes result in different advantages and disadvantages. They are both cheaper than isolated enzymes. Both show improved lifetimes in biosensors. The enzyme activity is stabilised in a more natural environment. They may be more stable to inhibition by solutes, pH and temperature changes. Their major disadvantage is some loss of selectivity as they often contain a mixture of enzymes. These may be of related type. Thus banana tissue, which was developed by Sidwell and Reichnitz (1985) and subsequently by Wang and Lin (1988) for the determination of dopamine, a catcholamine found in the brain and containing a complex of polyphenolases which catalyse the oxidation of polyphenolic compounds, has been found by ourselves (Eggins, 1994) to be equally effective for the determination of catechol itself and for flavanols, which are a type of catechol found as flavourings in beers and wines. This lowered selectivity can be turned to advantage in some sorts of analysis. Summarising their advantages and disadvantages:

Advantages
(i)   The enzyme(s) is maintained in its natural environment.
(ii) The enzyme activity is stabilised.
(iii) They sometimes work when purified enzymes fail.
(iv) They are much less expensive than purified enzymes.

Disadvantage
There may be interfering processes, i.e. there is some loss of selectivity.

2.4.1 Examples

1. The first example was described by Rechnitz (1978). Arginine is broken down by bovine liver to urea and ornithine. The urea is then determined with a urease potentiometric biosensor.

$$\text{arginine} \xrightarrow{\text{bovine liver}} \text{urea} + \text{ornithine}$$

$$\text{urea} + 2H_2O \xrightarrow{\text{urease}} 2NH_4^+ + HCO_3^-$$

2. Another amino acid, glutamine is broken down by pork liver to ammonia and glutamic acid. The ammonia is then measured using an ammonia ion-selective electrode.

$$\text{glutamine} + H_2O \xrightarrow{\text{pork liver}} NH_3 + \text{glutamate}$$

This assay has been studied using several different biological materials and a useful comparison has been made between them (see Table 2.3).

3. As already mentioned, banana tissue (and also aubergine, apple, cucumber and mushroom tissue) contains polyphenolases which catalyse the oxidation of molecules containing catechol groups to the corresponding *o*-quinone. Sidwell and Rechnitz (1985) evolved a biosensor for dopamine based on the banana, the 'bananatrode', using an oxygen electrode to follow the assay. This was simplified by Wang and Lin (1988), who incorporated the banana tissue into a carbon paste electrode. Detection was then by voltammetric reduction of the quinone. The electrode is shown in Figure 2.2 (see also Chapter 1).

Further examples of the use of tissue material are given in Table 2.1 (Arnold and Rechnitz, 1987).

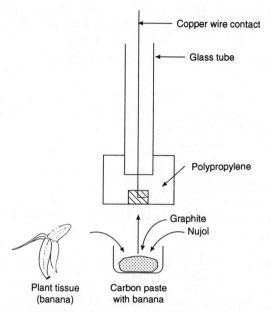

Figure 2.2 The banana electrode (cf. Chapter 1)

Table 2.1 Biosensors based on tissue and related materials

Substrate	Biocatalytic material	Sensing element
Glutamine	Porcine kidney cells	NH_3
Adenosine	Mouse small-intestine mucosal cells	NH_3
AMP	Rabbit muscle	NH_3
Guanine	Rabbit liver	NH_3
Hydrogen peroxide	Bovine liver	O_2
Glutamate	Yellow squash	CO_2
Pyruvate	Corn kernel	CO_2
Urea	Jack bean meal	NH_3
Phosphate/fluoride	Potato tuber/GOD	O_2
Dopamine	Banana pulp	O_2
Tyrosine	Sugar beet	O_2
Cysteine	Cucumber leaf	NH_3
Glutamine	Porcine kidney mitochondria	NH_3

2.5 MICROORGANISMS

Microorganisms play an important part in many biotechnological processes in industry, in fields such as brewing, pharmaceutical synthesis, food manufacture, waste water treatment and energy production. Many biosensors

based on microorganisms immobilised on a transducer have been developed
to assist with the monitoring of these processes and others.

Microorganisms can assimilate organic compounds, resulting in a change in
respiration activity, and can produce electroactive metabolites.

Advantages
  (i)   They are a cheaper source of enzymes than isolated enzymes.
 (ii)   They are less sensitive to inhibition by solutes and more tolerant of pH
        changes and temperature changes.
(iii)   They have longer lifetimes.

Disadvantages
  (i)   They sometimes have longer response times.
 (ii)   They have longer recovery times.
(iii)   Like tissues, they often contain many enzymes and so may have less
        selectivity.

2.5.1 Examples

1. In the fermentation of cane molasses, which contains various sugars, it is
important to determine the total assimible sugars in the broth. This can be
done using whole cells of *Brevibacterium lactofermentum* immobilised on an
oxygen electrode.

2. A glucose sensor was made using *Pseudomonas fluorescens* immobilised
on an oxygen electrode. The sensor structure is shown in Figure 2.3.

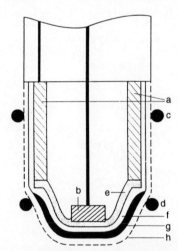

Figure 2.3 A glucose sensor using bacteria. Scheme of the microbial electrode for total
assimilable sugars. a, Anode; b, platinum cathode; c, d, rubber rings; e, electrolyte gel; f,
membrane; g, microorganisms retained on nylon net; h, Cellophane membrane.
Reproduced by permission of the Open University Press from Hall (1990)

3. An interesting biosensor pair is represented by the use of *Trichosporon brassicae* with an oxygen sensor for ethanol or for acetic acid. For the determination of acetic acid, the pH is kept well below the pK_a of acetic acid (4.75), as acetate ions cannot pass through the membrane. There was no response to volatile compounds such as formic acid and methanol, or to nutrients such as phosphate or glucose. There is interference from propanoic acid, butanoic acid and ethanol, but these are not usually present in fermentations of interest.

In the alcohol mode, the pH is kept above 6 to ensure that acetic acid, propanoic acid, etc., are in their ionised state and cannot pass through the membrane. This sensor is highly selective and has been developed commercially in Japan.

Some other microorganism based biosensors are shown in Table 2.2. taken from Karube (1987).

Table 2.2   Biosensors based on microorganisms

Sensor	Microorganism	Device/ probe	Response min	Range/ mg dm^{-3}
Assimilable sugars	*Brevibacterium lactofermentum*	O_2	10	10–200
Glucose	*Pseudomonas fluorescens*	O_2	10	2–20
Acetic acid	*Trichosporon brassicae*	O_2	10	3–60
Ethanol	*Trichosporon brassicae*	O_2	10	2–25
Methanol	Unidentified bacteria	O_2	10	5–20
Formic acid	*Citrobacter freundi*	Fuel cell	30	10–10^3
Methane	*Methylomonas flagellata*	O_2	2	0–7
Glutamic acid	*Escherichia coli*	CO_2	5	10–100
Cephalosporin	*Citrobacter freundi*	pH	10	100–500
BOD	*Trichosporon cutaneum*	O_2	15	3–60
Lysine	*Escherichia coli*	CO_2	5	10–100
Ammonia	Nitrifying bacteria	O_2	10	0.05–1
Nitrogen dioxide	Nitrifying bacteria	O_2	3	0.51–255
Nicotinic acid	*Lactobacillus arabinosis*	pH	60	10^{-5}–5
Vitamin B$_1$	*Lactobacillus fermenti*	Fuel cell	360	10^{-3}–10^{-2}

Reproduced by permission of Oxford University Press from Arnold and Rechnitz (1987).

Table 2.3, reproduced from Arnold and Rechnitz (1987), gives a classical comparison of relative performance criteria for the determination of glumtamine hydrolysis by glutaminase from pork liver, either as isolated enzyme, in pork liver tissue, in mitochondria or in a bacterial preparation.

$$\text{glutamine} + H_2O \xrightarrow{\text{glutaminase}} \text{glutamate} + NH_3$$

Table 2.3   Response characteristics of glutamine biosensors

Parameter	Enzyme	Mitochondria	Bacteria	Tissue
Slope/mV per decade	33–41	53	49	50
Detection limit/M	6.0×10^{-5}	2.2×10^{-5}	5.6×10^{-5}	2.0×10^{-5}
Linear range/mM	0.15–3.3	0.11–5.5	0.1–10	0.064–5.2
Response time/min	4–5	6–7	5	5–7
Lifetime/days	1	10	20	30

Reproduced by permission of Oxford University Press from Arnold and Rechnitz (1987).

The reaction is monitored potentiometrically with an ammonia ion-selective electrode.

One can see that the sensitivity (slope in mV per decade) is poorest with the pure enzyme and higher with the other preparations, but never quite reaching the Nernstian value of 59 mV per decade. The responses in terms of detection limits and linear range are similar, with pork liver tissue showing the best response. The response time is long for all preparations, but worst for the tissue. However, the working lifetime is dramatically improved from only 1 day with the pure enzyme to 30 days with the tissue extract.

2.6 MITOCHONDRIA

These sub-cellular multi-enzyme particles can be effective biocatalytic components. They can sometimes be useful in improving sensor response and selectivity when the entire tissue lacks the necessary properties. Table 2.3 shows the relative performance characteristics of a mitochondrial biosensor for glutamine.

2.7 ANTIBODIES

Organisms develop antibodies (Ab) which are proteins that can bind with an invading antigen (Ag) and remove it from harm (Figure 2.4).

$$Ab + Ag \rightleftharpoons Ab \cdot Ag$$

The affinity constant, $K = [Ag \cdot Ab]/[Ag][Ab]$, is usually about 10^6. For a fixed concentration of antibody, the ratio of the free to bound antigen, $[Ag]/[Ag \cdot Ab]$, at equilibrium is quantitatively related to the total amount of ligand, i.e. $[Ag] + [Ag \cdot Ab]$. If a fixed amount of Ag is added to the assay, the unknown concentration of the original Ag can be determined.

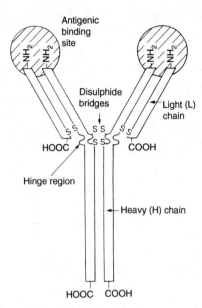

Figure 2.4 Scheme for a typical antibody composed of two heavy chains and two light chains. Reproduced by permission of the Open University Press from Hall (1990)

Unknown Ab can be determined by using labelled antibodies. Labelling may be done with radioisotopes, enzymes, red cells, fluorescent probes, chemiluminescent probes or metal tags. Actually, the label may be on either the Ab or the Ag. In biosensors the emphasis has been on labelling with enzymes.

Advantages
(i)   They are very selective (they can be too selective between strains).
(ii) They are ultra-sensitive.
(iii) They bind very powerfully.

Disadvantages
There is no catalytic effect.

Antibodies have long been used in immunoassays. They bind even more powerfully and specifically to the corresponding antigen than enzymes do to their substrates. In fact, they can be too selective between different strains of the same material. They are ultra-sensitive, although they lack the catalytic activity of enzymes. They are often used in a labelled form (see above).

One can classify enzyme-linked immunoassays with the electroanalytical method used, as follows.

2.7.1   Amperometric Assays Based on the Clarke Oxygen Electrode

An example is the determination of human chorioic gonadotropin (hCG) using catalase-labelled hCG. An antibody was immobilized on a cellulose membrane, which was then placed over the PTFE membrane of the oxygen electrode. Figure 2.5 shows the arrangement. Both labelled and unlabelled hCG were allowed to compete for the antibody in the membrane. The membrane was then washed to remove bound from free hCG, and the electrode was exposed to hydrogen peroxide solution, which produced oxygen with the catalase. The oxygen was then measured at the electrode.

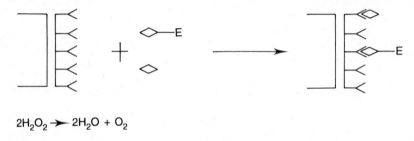

$$2H_2O_2 \rightarrow 2H_2O + O_2$$

Figure 2.5   An hCG amperometric sensor. Reproduced by permission of Oxford University Press from Green (1987)

2.7.2   Amperometric Enzyme-linked Immunoassays

An example is human serum orosomucoid (hso, a carcinogen) labelled with alkaline phosphatase (AP). There is a competitive reaction between the antibody and hso immoblised on the surface of a cuvette and enzyme-labelled hso. This is washed and the substrate phenyl phosphate is added:
phenyl phosphate + H_2O → phosphate + phenol
The phenol concentration is measured by electrochemical oxidation at a carbon paste electrode.

2.7.3   Amperometric Immunoassays Labelled with an Electroactive Species

An example is the labelling of morphine with ferrocene. This was electrochemically oxidised in the presence and absence of the antibody against morphine. This assay also worked with codeine.

2.7.4   Potentiometric Immunoassay

The development of biosensors based on antibodies using potentiometric transducers was slow and had many problems. However, several satisfactory examples have been demonstrated. One such is for the determination of oestradiol-17β, using an iodide-selective electrode. The anti-oestradiol-17β was immoblised on a gelatine membrane on the surface of an iodide ISE. The assay was performed by competition between peroxidase-labelled antigen and sample antigen. This is shown in Figure 2.6.

The amount of sample antigen is inversely proportional to the labelled antigen. The labelled antigen is determined by adding H_2O_2 and iodide, which are then converted into iodine in the presence of the peroxidase. The ISE then measures the iodine and a calibration graph of e.m.f. versus log[oestradiol] is obtained (Figure 2.6).

Another example is the assay of digoxin using a carbon dioxide electrode, as shown in Figure 2.7.

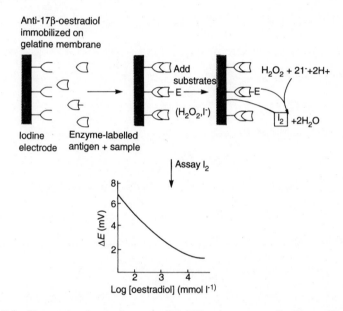

Figure 2.6 Determination of oestradiol-17β via an antibody-modified iodide electrode. Reproduced by permission of the Open University Press from Hall (1990)

2.7.5 Example

A dramatic example is the use of antibodies in the determination of TNT, using a photometric method. The TNT is incubated with the immoblised Sepharose antibody to which it binds. TNT labelled with glucose-6-phosphate dehydrogenase is added to form G-6-PDH–TNT. The Sepharose is washed to remove unbound TNT, then the assay is carried out with NAD[+], FMN and luciferase. Detection is photometric measurement of the emitted light.

$$\text{glucose-6-phosphate} + \text{NAD}^+ \xrightarrow{\text{G-6-PDH–TNT}}$$
$$\text{glucono-6}'\text{-lactone-6-phosphate} + \text{NADH} + \text{H}^+$$

$$\text{NADH} + \text{H}^+ + \text{FMN} \xrightarrow{\text{oxidoreductase}} \text{NAD}^+ + \text{FMNH}_2 \xrightarrow{\text{luciferase}} h\nu$$

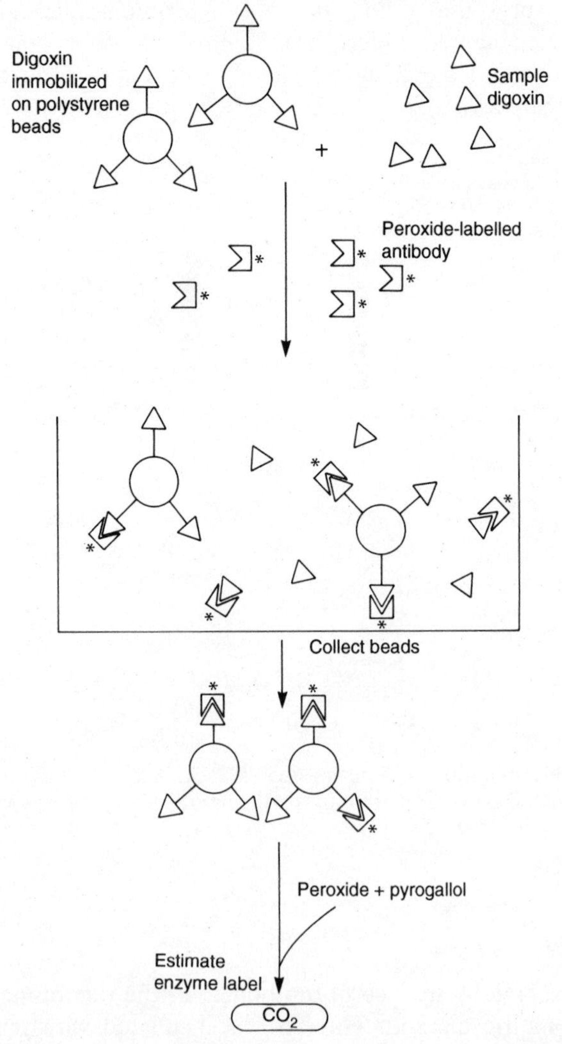

Figure 2.7 A potentiometic immunoassay: enzyme-linked immunoassay of digoxin via carbon dioxide electrode. Reproduced by permission of the Open University Press from Hall (1990)

This is an extremely sensitive assay which can detect down to 10^{-18} mol (1 amol) of TNT.

2.8 NUCLEIC ACIDS

Nucleic acids operate in many ways like antibodies. The specific base pairing between strands of nucleic acid gives rise to the genetic code which determines the replicating characteristics of all parts of living cells and thus the inherited characteristics of individual members of a species. There is thus a genetically coded nucleic acid for each individual molecule created in and by a living cell, including proteins and hence enzymes.

DNA probes can be used to detect genetic diseases, cancers and viral infections. They are used either in the 'short' synthetic form or the 'long' form produced by cloning. As with antibodies, a DNA assay often involves the addition of labelled DNA to the assay. The labelling may be radioactive, photometric, enzymic or electroactive, giving scope for a range of biosensor types.

With genetic and protein engineering, another use for DNA in developing sensors is possible. A number of approaches have been envisioned, as follows:

1. *Improvement of enzyme yield.* Some useful enzymes are present in such small quantities that they are difficult to isolate. An example is glucose dehydrogenase (GDH), which is of potential use in glucose biosensors which would have no requirement for oxygen (or other oxidant), unlike glucose oxidase. One source of GDH is the bacterium *Acinetobacter calcoaceticus*, but it is present only at low levels. This level can be greatly enhanced using cloning techniques, which effectively duplicate the gene encoding GDH, producing up to 50 copies of the plasmid vector per cell and consequently large amounts of the enzyme. The details of this involve a knowledge of genetic engineering techniques.

2. *Improvement of enzyme properties.* Changes in the enzyme itself can potentially be made either directly by protein engineering or by modifying the replicating gene using genetic engineering. Modifications might (a) improve the turnover number of the enzyme, (b) change the pH dependence, (c) change the linear response to substrate concentration, (d) improve the stability during storage and operation, (e) reduce the susceptibility to interfering substances, (f) widen or narrow substrate specificity and (g) change the cofactor requirement. Considerable advances have been made in these areas, although there is not yet much regular application.

2.9 RECEPTORS (Leech, 1994)

Much of the research into receptors has been with neuroreceptors and their recognition of neurotransmitters, neurotransmitter antagonists and neurotoxins. Work is only just beginning as far as application to biosensors is concerned, and it is difficult to present any coherent classification of types of receptors. However, we can give a series of interesting examples of applications and potential applications.

Neurotransmitter and hormonal receptors are the body's own biosensors. An example, already referred to in Chapter 1, is the olfactory membrane in the nose.

Binding of a ligand (an agonist) to the receptor triggers an amplified physiological response such as (i) ion channel opening, (ii) second messenger systems and (iii) activation of enzymes. These biological receptors tend to have an affinity for a range of structurally related compounds rather than one specific analyte, which is often an attractive feature for use in biosensors. Often they are used with labelled materials, originally with radioactive labelled ligands but now also with fluorescent- or enzyme-labelled ligands. Alternatively, the label could be on the receptor.

They may be broadly grouped into intact receptor-based biosensors and isolated receptor-based biosensors.

In the former group, one of the most interesting experiments, resulting in a so called *prototype chemoreceptor* (Belli and Rechnitz, 1986), used excised crustacean olfactory structures for neuronal sensing of chemoreceptor ligands such as amino acids, sex hormones and pyridine-based chemicals. The binding of the ligand to the corresponding chemoreceptor triggers the firing of an action potential (AP) in the nerve which can be detected by a glass microelectrode outside the cell.

A system using the giant axions of the crayfish was used to detect local anaesthetics, which can bind to the ion channel of the voltage-gated sodium channel receptor responsible for AP propogation, thus inhibiting the AP conduction.

This system can also be applied to neuromodulator drugs or toxins, including antidepressants, narcotics, alcohols and venoms. They all affect AP conduction. Unfortunately, such systems have a lifetime of only 4–8 h.

One of the most studied of the isolated receptors is nicotinic acetylcholine. It has been used with several transducers, including ISFET and capacitance types. However, the most successful has been the use of a fibre-optic evanescent wave with fluorescent-labelled ligands (see Chapter 5). This has been used with FITC-labelled α-conotoxin. The sensor had a lifetime of 30 days.

There are many other interesting aspects of receptors to be investigated before successful biosensors can be developed.

2.10 REFERENCES

M. A. Arnold and G. A. Rechnitz (1987) 'Biosensors based on plant and animal tissues', in A. P. F. Turner, I. Karube and G. S. Wilson (Eds), *Biosensors: Fundamentals and Applications,* Oxford University Press, Oxford, Chap. 3, pp. 30–59.

S. L. Belli and G. A. Rechnitz (1986) *Anal. Lett.* **19**, 403.

A. E. Cass and E. Kenny (1987) 'Protein engineering and its potential application to biosensors', in A. P. F. Turner, I. Karube and G. S. Wilson (Eds), *Biosensors: Fundamentals and Applications*, Oxford University Press, Oxford, Chap. 8, pp. 113–132.

B. R. Eggins (1994) 'Recent advances in the fabrication and application of biosensors', in *Proceedings of Conference on Analytical Advances in the Biosciences, University of Ulster, 23–24 June 1994*, p. 16.

T. D. Gibson and R. J. Woodward (1992) 'Protein stabilisation in biosensor systems', in P. F. Eldman and J. Wang (Eds), *Biosensors and Chemical Sensors*, American Chemical Society, Washington, DC, Chap. 5, pp. 40–55.

T. D. Gibson, J. N. Hulbert, S. M. Parker, J. R. Woodward and I. J. Higgins (1992) 'Extended shelf life of enzyme-based biosensors using a novel stabilisation system', *Biosensors Bioelectron.,* **7**, 701.

M. J. Green (1987) 'New approaches to electrochemical immunoassays', in A. P. F. Turner, I. Karube and G. S. Wilson (Eds), *Biosensors: Fundamentals and Applications*, Oxford University Press, Oxford, Chap. 4, pp. 60–70.

E. A. H. Hall (1990) 'The biomolecule reviewed', in *Biosensors*, Open University Press, Milton Keynes, Chap. 2.

I. Karube (1987) 'Micro-organism based sensors', in A. P. F. Turner, I. Karube and G. S. Wilson (Eds), *'Biosensors: Fundamentals and Applications*, Oxford University Press, Oxford, Chap. 2, pp. 13–29.

S. S. Kuan and G. G. Guilbault (1987) 'Ion selective electrodes and biosensors based on ISE's', in A. P. F. Turner, I. Karube and G. S. Wilson (Eds), *Biosensors: Fundamentals and Applications*, Oxford University Press, Oxford, Chap. 9, pp. 135–152.

D. Leech (1994) 'Affinity biosensors', *Chem. Soc. Rev.*, 205.

G. A. Rechnitz (1978) 'Biochemical electrode uses tissue slice', *Chem. Eng. News*, **56**, 16.

K. R. Rogers and J. N. Lin (1992) 'Biosensors for environmental monitoring', *Biosensors Bioelectron.*, **7**, 317.

F. W. Scheller, D. Pfeiffer, F. Schubert, R. Renneberg and D. Kirsten (1987) 'Application of enzyme-based amperometric biosensors to the analysis of "real" samples', in A. P. F. Turner, I. Karube and G. S. Wilson (Eds), *Biosensors: Fundamentals and Applications*, Oxford University Press, Oxford, Chap. 18, pp. 315–346.

J. S. Sidwell and G. A. Rechnitz (1985) '"Bananatrode"—an electrochemical biosensor for dopamine', *Biotechnol. Lett.*, **7**, 419.

P. Vadgama and P. W. Crump (1992) 'Biosensors: recent trends', *Analyst*, **117**, 1657.

J. Wang and M. S. Lin (1988) 'Mixed plant tissue–carbon paste electrode'. *Anal. Chem.*, **60**, 1545.

P. G. Warner (1987) 'Genetic engineering', in A. P. F. Turner, I. Karube and G. S. Wilson (Eds), *Biosensors: Fundamentals and Applications*, Oxford University Press, Oxford, Chap. 7, pp. 100–112.

S. G. Weber and A. Webers (1993) 'Biosensor calibration. *In situ* recalibration of competitive binding sensors', *Anal Chem.*, **65**, 223.

R. Wilson and A. P. F. Turner (1992) 'Glucose oxidase: an ideal enzyme', *Biosensors Bioelectron.* **7**, 165.

Chapter 3

Immobilisation of Biological Component

3.1 INTRODUCTION

In order to make a viable biosensor, the biological component has to be properly attached to the transducer. This process is known as *immobilisation*. There are five regular methods of doing this, as follows.

(i) *Adsorption*

This is the simplest and involves minimal preparation. However, the bonding is weak and this method is only suitable for exploratory work over a short time-span.

(ii) *Microencapsulation*

This was the method used in the early biosensors. The *biomaterial* is held in place behind a membrane, giving close contact between the biomaterial and the transducer. It is adaptable. It does not interfere with the reliability of the enzyme. It limits contamination and biodegredation. It is stable towards changes in temperature, pH, ionic strength and chemical composition. It can be permeable to some materials, e.g. small molecules, gas molecules and electrons.

(iii) *Entrapment*

The biomaterial is mixed with monomer solution, which is then polymerised to a gel, trapping the biomaterial. Unfortunately, this can cause barriers to the diffusion of substrate, thus slowing the reaction. It can also result in loss of bioactivity through pores in the gel. This can be counteracted by cross-linking (see below). The most commonly used gel is polyacrylamide, although starch gels, nylon and silastic gels have been used. Conducting polymers such as polypyrroles are particularly useful with electrodes.

(iv) *Cross-linking*

In this method, the biomaterial is chemically bonded to solid supports or to another supporting material such as a gel. Bifunctional reagents such as glutaraldehyde are used. Again there is some diffusion limitation and there can be damage to the biomaterial. Also, the mechanical strength is poor. It is a useful method to stabilise adsorbed biomaterials.

(v) *Covalent bonding*

This involves a carefully designed bond between a functional group in the biomaterial to the support matrix. Nucleophilic groups in the amino acids of the biomaterials, which are not essential for the catalytic action of, say, an enzyme are suitable. Many examples are known. We shall illustrate the technique with one example. A carboxyl group on the support is reacted with a carbodiimide. This then couples with an amine group on the biomaterial to form an amide bond between the support and the enzyme, as shown in Figure 3.1.

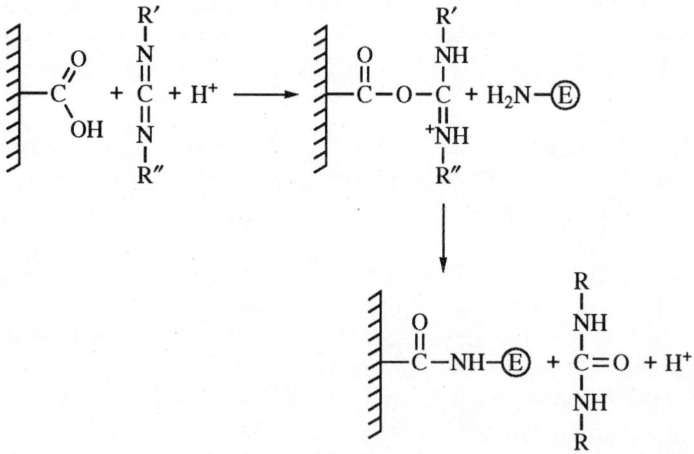

Figure 3.1   Covalent bonding of an enzyme to a transducer via a carbodiimide

The reaction must work at low temperature, low ionic strength, and neutral pH. The enzyme will not be lost from the biosensor during use with this method.

Overall, the lifetime of the biosensor is greatly enhanced by proper immobilisation. Typical lifetimes for the same biosensor, in which different methods of immobilisation are used, are as follows:

adsorption	1 day
membrane entrapment	1 week
physical entrapment	3–4 weeks
covalent entrapment	4–14 months

3.2 ADSORPTION

Many substances adsorb enzymes on their surfaces, e.g. alumina, charcoal, clay, cellulose, kaolin, silica gel, glass and collagen. No reagents are required, there is no clean-up step and there is less disruption to the enzymes.

Adsorption can roughly be divided into two classes: physical adsorption (physisorption) and chemical adsorption (chemisorption). Physisorption is usually weak and occurs via the formation of van der Waals bonds, occasionally with hydrogen bonds or charge-transfer forces. Chemisorption is much stronger and involves the formation of covalent bonds.

Several model equations are used to describe adsorption, but the most generally used is the Langmuir adsorption isotherm. This is derived by kinetic considerations and relates the fraction of the surface covered (θ) by adsorbent with kinetic parameters:

$$\text{rate of adsorption} = K_a p_a N(1 - \theta)$$

$$\text{rate of desorption} = K_d N \theta$$

At equilibrium the two rates are equal, so

$$\theta = K p_a / (1 + K p_a)$$

where p_a = pressure of adsorbent, k_a = rate constant for adsorption, k_d = rate constant for desorption and $K = k_a/k_d$.

Adsorbed biomaterial is very susceptible to changes in pH, temperature, ionic strength and the substrate. However, the method is satisfactory for short-term investigations.

3.3 MICROENCAPSULATION

In this method, an inert membrane is used to trap the biomaterial on the transducer. It was the method used with the first glucose biosensor on the oxygen electrode.

The advantages are as follows:

(i)   There is close attachment between the biomaterial and the transducer.
(ii) It is very adaptable.
(iii) It is very reliable.

(iv)   The reliability of the biomaterial (enzyme) is maintained as follows:
    (a)   a high degree of specificity is maintained;
    (b)   there is good stability to changes in temperature, pH, ionic strength, $E°$ and substrate concentration;
    (c)   it is an inbuilt device to limit contamination and biodegradation;
    (d)   if used with a patient, it avoids infection.
 (v)   There is always the option of bonding the biological component to the sensor via molecules that conduct electrons, such as polypyrrole.

The main types of membranes used are the following: *cellulose acetate* (dialysis membrane), which excludes proteins and slows the transportation of interfering species such as ascorbate; *polycarbonate (Nuclepore)*, a synthetic material which is non-permselective; *collagen*, a natural protein; and *PTFE*, polytetrafluoroethylene (often known under the trade-name Teflon), a synthetic polymer, which is selectively permeable to gases such as oxygen. Other materials sometimes used are *Nafion* and *polyurethane*.

3.4   ENTRAPMENT

A polymeric gel is prepared in a solution containing the biomaterial. The enzyme is thus trapped within the gel matrix.

The most common polymer used is polyacrylamide. It is prepared by copolymerisation of acrylamide with *N,N'*-methylenebisacrylamide. Polymerisation can be effected by UV radiation in the presence of vitamin B_1 as photosensitiser. Other materials used are starch gels, nylon, silastic gels and conducting polymers such as polypyrrole.

The problems are the following:

 (i)   Large barriers are created, inhibiting the diffusion of the substrate, which slows the reaction and hence the response time of the sensor.
(ii)   There is loss of enzyme activity through the pores in the gel. This problem may be overcome by cross-linking, with, e.g., glutaraldehyde.

3.5   CROSS-LINKING

This method uses bifunctional agents to bind the biomaterial to solid supports. It is a useful method to stabilise adsorbed enzymes.
Disadvantages are as follows:

  (i)   It causes damage to the enzyme.
 (ii)   It limits diffusion of the substrate.
(iii)   There is poor rigidity (mechanical strength).

Figure 3.2 shows the most commonly used materials, which are glutaraldehyde (which will react with lysine amino acid residues in the enzyme), hexamethylene diisocyanate and 1,5-dinitro-2,4-difluorobenzene.

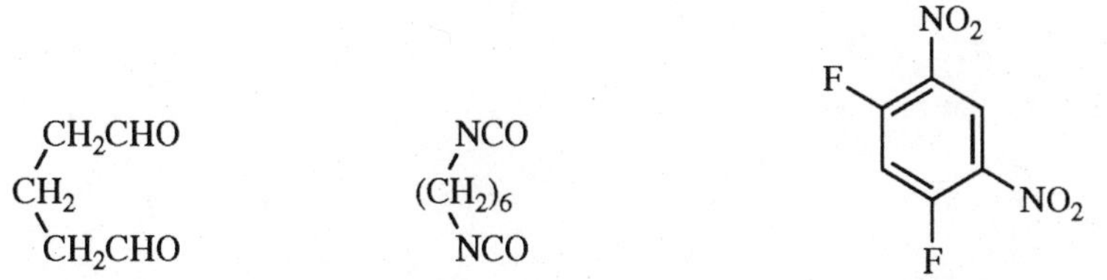

CH$_2$CHO / CH$_2$ \ CH$_2$CHO	NCO / (CH$_2$)$_6$ \ NCO	
Glutaraldehyde	Hexamethyl diisocyanate	1,5-dinitro-2,4-difluorobenzene

Figure 3.2 Some molecules used for cross-linking

3.6 COVALENT BONDING

Some functional groups which are not essential for the catalytic activity of an enzyme can be covalently bonded to the support matrix (transducer or membrane). This method uses nucleophilic groups for coupling such as NH$_2$, CO$_2$H, OH, C$_6$H$_4$OH, SH and imidazole.

Figure 3.3 shows the most common examples of the reactions used.

Reactions need to be performed under mild conditions—low temperature, low ionic strength and pH in the physiological range.

The advantage is that the enzyme will not be released during use. In order to protect the active site, the reaction is often carried out in the presence of the substrate.

In practice, it is unusual for only one of these methods to be used at a time, as some of the examples below will illustrate.

3.7 MODIFIED ELECTRODES

In parallel with the development of biosensors has been the development of so-called modified electrodes. In general, the modification has been in the form of polymer coatings on the electrodes. These give many new possibilities for their use as sensors and biosensors. The mode of attachment of biomaterial to these electrodes can involve various of the above methods. Because of their importance, it is pertinent to give some examples of such electrodes and their applications in biosensors. In fact, one could say that electrochemically transduced biosensors are all forms of modified electrodes.

In recent years, a new type of selectivity has been introduced by modifying

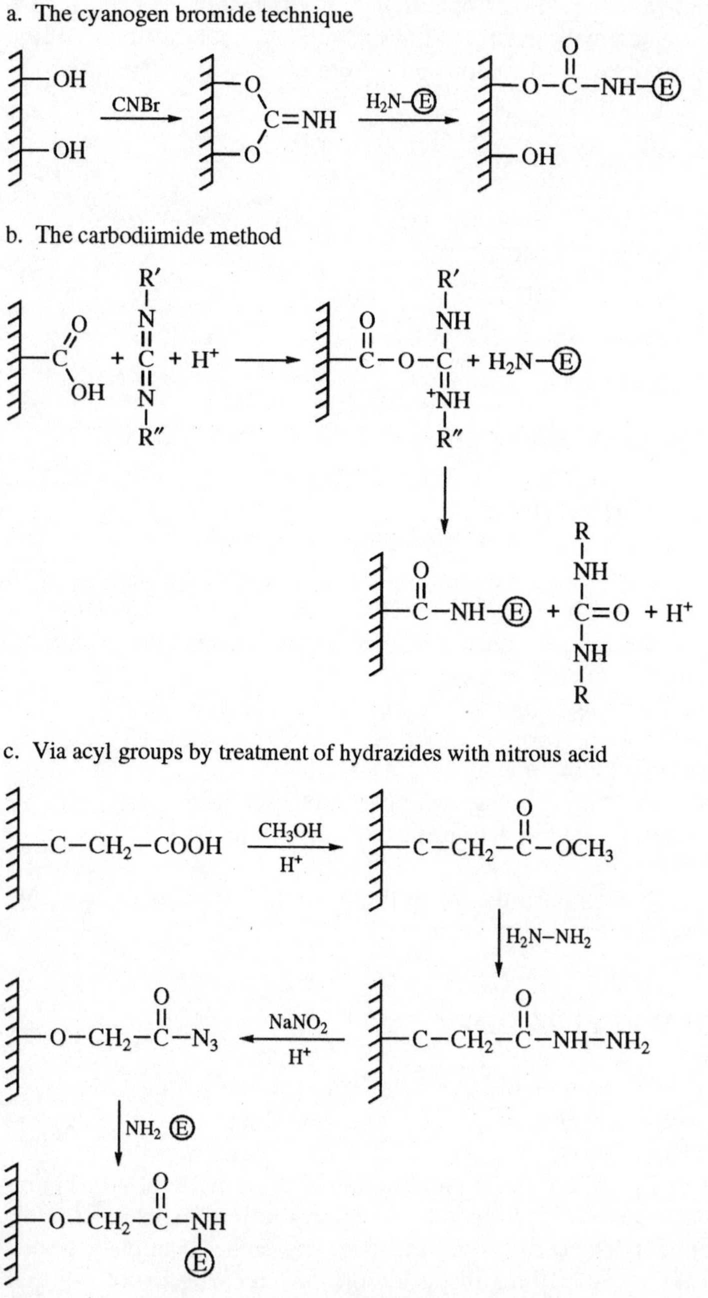
a. The cyanogen bromide technique
CNBr
H_2N-E
b. The carbodiimide method
c. Via acyl groups by treatment of hydrazides with nitrous acid
CH_3OH
H^+
H_2N-NH_2
$NaNO_2$
H^+
NH_2 E

Figure 3.3 Some common reactions used for covalent bonding. Reproduced by permission of the Open University Press from Hall (1990)

d. Coupling using cyanuric chloride

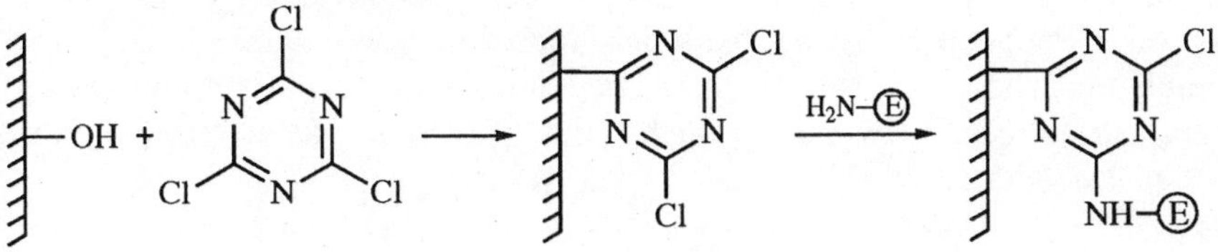

e. Coupling through diazonium groups from aromatic amino groups

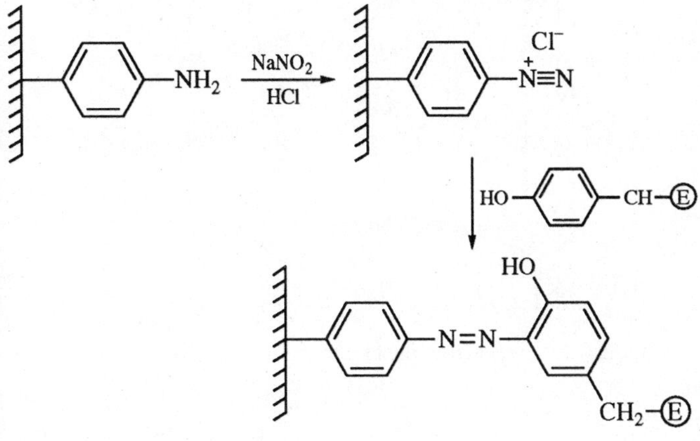

f. Coupling via thiol groups

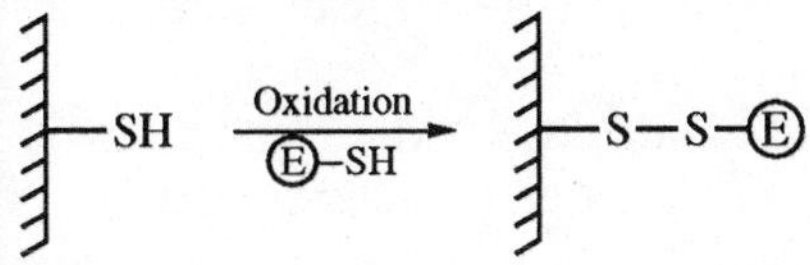

Figure 3.3 *(continued)*

the electrode surface. This can amount to changing the structure of the electrode material at the surface or to coating the surface with a new material which can have a highly selective effect. If the surface-modifying material is biological in nature, we have a *biosensor*. However, if the surface is modified before addition of the biomaterial, it may facilitate the immobilisation of the latter on the surface.

3.7.1 Modified Carbon Paste Electrodes

One of the simplest types of modified electrode is the modified carbon paste electrode. The carbon paste electrode (CPE) was invented in the 1960s (Adams, 1969) for studying oxidation reactions when mercury is not a suitable electrode material. It consists of a simple mixture of graphite power with Nujol to form a stiff paste, which is then placed into an electrode holder such as shown in Figure 3.4.

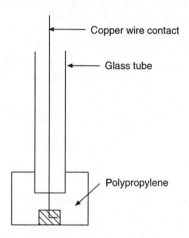

Figure 3.4 A carbon paste electrode

It is simplicity itself to mix a modifying component in with the paste. The component could be itself electroactive, as with ferrocene, used in the glucose biosensor, or maybe a complexing agent which can extract an electroactive analyte into the surface of the paste.

Another biosensor application is the 'bananatrode', in which banana pulp is mixed with the paste and used to determine dopamine and other catechols (see Chapters 2 and 9) (Wang and Lin, 1988).

3.7.2 Polymer Electrodes (Lyons, 1991; Teasdale and Wallace 1993)

A great deal of research has been carried out in recent years into modifying electrode surfaces by coating them with different types of polymer. These have been mainly of three types: conducting polymers, ion-exchange polymers and redox polymers.

Often chemical groups are attached to these coatings (or incorporated in their structure) to introduce particular electrochemical effects.

The applications have mainly been with voltammetric processes, but they have also been used in both potentiometric and conductimetric applications.

3.7.3 Electronically Conducting Polymers (Lyons, 1991; Teasdale and Wallace, 1993)

The most studied conducting polymers have been polyacetylene, polypyrrole, polyaniline and polythiophene. They are easily prepared by electrochemically oxidising the substrate on the electrode surface. The solvent used, and more particularly the counter anion in the solution, have a major effect on the properties of the polymer, and in particular on its selectivity characteristics for use in sensors. We shall give most attention to polypyrrole as it is the most versatile and the most studied. However, similar principles apply to other conducting polymers.

A solution of pyrrole in aqueous 0.1 M KCl can be oxidised at +0.8 V vs SCE at a Pt or glassy carbon electrode to form a layer of polypyrrole on the electrode surface according to the sequence shown in Figure 3.5.

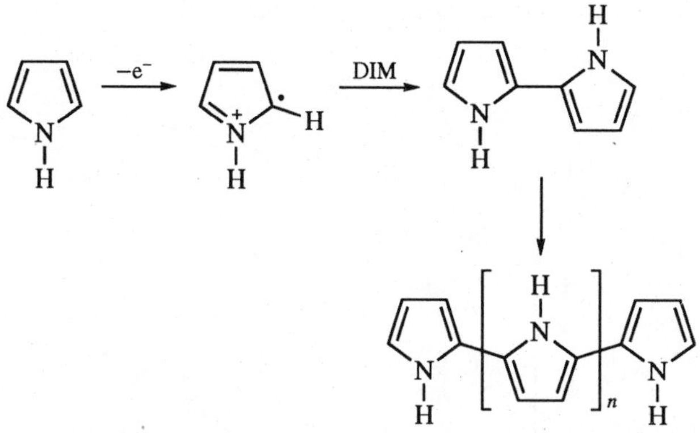

Figure 3.5 Oxidative polymerisation of pyrrole

Chemical methods of preparation can be used such as oxidation by halogens, SbF_5, AsF_5 or $FeCl_3$. The electrochemical method can be varied by changing the solvent and more interestingly the electrolyte. The most important electrochemical property of these conducting polymers is their redox switching. The polymers are electroactive and at certain potentials, as shown in the cyclic voltammogram in Figure 3.6, reduction can occur from the electronically conducting state to the non-electronically conducting form (but still *ionically conducting*):

$$PPY + A^-_{(soln.)} \rightleftharpoons PPY^+ + A^-_{(poly.)}$$

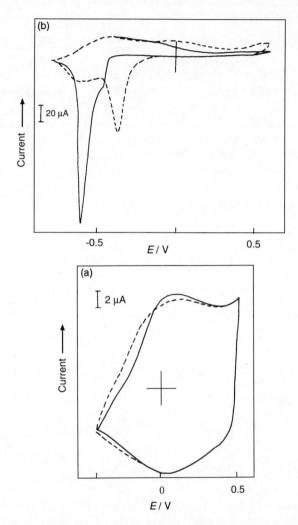

Figure 3.6 Cyclic voltammograms of polypyrrole with different counter anions: (a) in chloride; (b) in dodecylbenzenesulphonate. Reproduced by permission of the Royal Society of Chemistry from Lyons *et al.* (1993)

Polyaniline has three different oxidation states and hence three switching modes, as shown in Figure 3.7. Such changes have a dramatic effect on the electrical resistance of the material, resulting in large changes in its molecular recognition capabilities.

A wide range of types of molecular interaction contribute to variations in

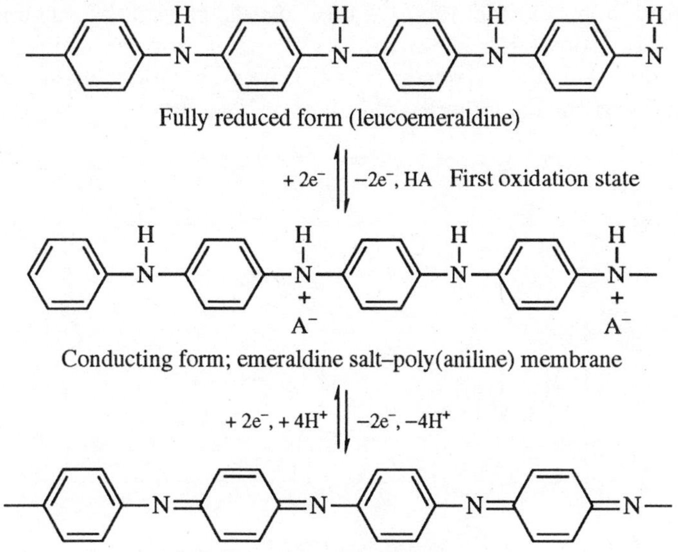

Fully reduced form (leucoemeraldine)

$+ 2e^-$ ⇅ $-2e^-$, HA First oxidation state

Conducting form; emeraldine salt–poly(aniline) membrane

$+ 2e^-, + 4H^+$ ⇅ $-2e^-, -4H^+$

Fully oxidised form (pernigraniline)

Figure 3.7 Oxidation of aniline to polyaniline showing switching modes. Reproduced by permission of the Royal Society of Chemistry from Teasdale and Wallace (1993)

molecular recognition. These are principally the following:

ion–ion;
ion–dipole;
dipole–dipole (H-bonding);
ion–induction;
dipole–induction;
dispersion.

Analyte recognition is generally by either (a) chelation and complexation, (b) enzyme–substrate interaction or (c) antibody–antigen interaction. These can be accomplished by incorporation of the appropriate group into the polymer matrix, such as EDTA in polypyrrole, or by incorporation of biorecognition elements such as glucose oxidase or urease into the polymer. Work has also been done with polynucleotides and with antibodies.

Conducting polymers are inherently ion-exchange materials, and the nature of the counter-ion can greatly modify the relative exchange properties of the polymer. Thus, with PPy–Cl^- the order of exchange is $BR^- > SCN^- > SO_4^{2-} > I^- > CrO_4^{2-}$, but with PPy–$ClO_4^{2-}$ the order is $SCN^- > Br^- > I^- > SO_4^{2-} > CrO_4^{2-}$.

If one uses a hydrophobic counter-ion such as dodecylsulphate or poly(vinylsulphonate), the ion-exchange character can be eliminated or modified. The polymer can also be transformed from an anion-exchange polymer into a cation exchanger as shown in the scheme and the accompanying cyclic voltammogram in Figure 3.8.

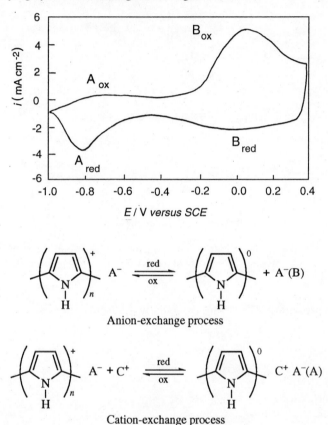

Anion-exchange process

Cation-exchange process

Figure 3.8 Pyrrole based anion–cation-exchange polymers. Top: cyclic voltammogram of polypyrrole at a microelectrode in $10^{-3}\,\text{mol}\,l^{-1}$ KCl. Reproduced by permission of the Royal Society of Chemistry from Lyons *et al.* (1993)

3.7.4 Ion-exchange Polymers (Espenscheid *et al.*, 1986)

Ionomer film-modified electrodes are another type of polymer-coated electrode. An ionomer is a linear- or branched-chain polymer containing covalently attached ionisable groups. They contain <10% of ionisable monomers and are not cross-linked.

They are not themselves electroactive, but they show unusually high ion-exchange selectivity for large hydrophobic cations, such as alkylamonium ions, compared with mono- and divalent inorganic ions. The main types are perfluorosulphonate polymers such as (a) Nafion (du Pont) and (b) PFASA (Dow);

(a) $[CF_3(CF_2)_n]_2 = CF—O—CF_2—CF(CF_3)—O—(CF_2)_2—SO_3H$
(b) $[CF_3(CF_2)_n]_2 = CF—O—(CF_2)_2—SO_3H$

The ionomer film will extract and preconcentrate large cations from an aqueous phase. Cations which have been studied include electroactive ions such as methyl viologen (MV^{2+}), ferrocenylmethyltrimethylammonium (FA^+) and ruthenium complexes such as $Ru(NH_3)_4^{3+}$ and $Ru(byp)_3^{2+}$. These can all be detected by voltammetry.

The interaction between the aqueous solution and the film can be expressed by the following equation:

$$O^{n+}(aq.) + nNa^+ (film) \rightarrow O^{n+} (film) + nNa^+(aq.)$$

This reaction can be quantified by the distribution coefficient, k_D, between water and the film, and $K_{O/Na}$, the equilibrium constant for this reaction. Values of k_D and $K_{O/Na}$ for these ions are given in Table 3.1. Some values of $K_{O/Na}$ for simple inorganic ions are $Cs^+ = 9.1$ and $Ba^{2+} = 30$.

Table 3.1 Ion-exchange distribution coefficients, k_D, and selectivity coefficients, $K_{O/Na}$. Reproduced by permission of the Royal Society of Chemistry from Espenscheid *et al.* (1986).

Cation	k_D	$K_{O/Na}$
$MV^{2+ a}$	7.9×10^5	1.5×10^4
$FA^{+ b}$	1.1×10^6	7.3×10^4
$Ru(NH_3)_6^{3+}$	2.5×10^6	3.7×10^4
$Ru(bpy)_3^{2+}$	2.1×10^7	5.7×10^6
$Ru(NH_3)_6^{2+}$	2.6×10^4	7.4×10^2

[a]Methyl viologen.
[b]Ferrocenylmethyltrimethylammonium.

From the analytical point of view, the procedure is similar to that used in anodic stripping voltammetry. As can be seen in Figure 3.9, concentrations of 10^{-8} M can easily be detected with CV currents of about 2–20 μA. The technique can be called *ion-exchange voltammetry*.

In order to establish a viable system, a number of factors need to be considered:

(i) film mass transport dynamics, which govern the speed of film/solution equilibration;

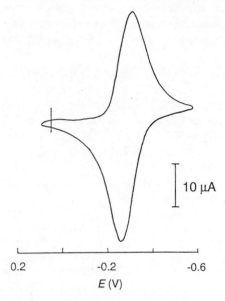

0.2   -0.2   -0.6

E (V)

Figure 3.9 Cyclic voltammogram of Nafion coated electrode in equilibrium with 2.78 × 10^{-8} M Ru(NH$_3$)$_6^{3+}$. Reproduced by permission of the Royal Society of Chemistry from Espenscheid *et al.* (1986)

(ii)   ion-exchange selectivity, which governs detection limits;
(iii) regeneration of the film for a new run;
(iv)   selectivity between analytes. Ionomers have only a very rudimentary selectivity, although they are better than the bare electrode. This factor needs to be further studied. Approaches may be to build selectivity into the membrane, as with chemically selective membranes used in potentiometry.

An interesting application of the technique is the *in vivo* determination of neurotransmitter substances such as dopamine in rat brains, in the presence of potentially interfering substances such as ascorbate. The former are protonated at physiological pH and will be extracted by the Nafion film, whereas ascorbate exists as the anion and is rejected.

3.7.5   Redox Polymers (Lyons, 1991)

A number of modified electrodes have been made with a redox group attached to the polymer coating. A popular approach has been to polymerise 4-vinylpyridine on the surface. This will then selectively coordinate with transition metal ions. Ruthenium and osmium ions have been particularly studied, as the resulting redox polymers are effective redox catalysts for other analytes.

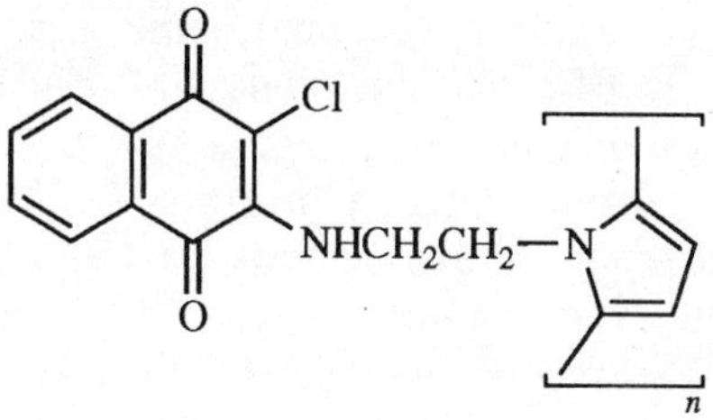

Figure 3.10 A redox polymer using a quinone attached to polypyrrole

A similar approach has been used with polypyrroles, poly-*N*-methylenepyrroles and polythiophenes (Grimshaw and Perera, 1990), using mainly covalently attached quinones as the redox group, as illustrated by the example in Figure 3.10.

Most of this work has been done with chlorinated benzoquinones and naphthoquinones, and a sulphonated anthraquinone, in order to keep the redox potential low. So far these polymers have been characterised, but not yet applied to sensors.

Another approach has been to polymerise amino acids containing built-in redox groups. Such polymerisations occur readily at moderate pHs. They have been developed with 4-nitrobenzoyl groups attached to poly-L-lysine, poly-L-ornithine and poly-L-glutamate (Abeysekera *et al.*, 1992). With an imidazole group attached to poly-L-ethylglutamate, an iron–porphyrin group was attached as the redox group via a trimethyleneamino group to make a cytochrome *c* analogue (Grimshaw and Grimshaw, 1990).

3.7.6 Phthalocyanines (Snow and Barger, 1989)

Phthalocyanines (Figure 3.11) are weak semiconductors, but this conductivity can be greatly increased by complexation with certain electron-acceptor molecules such as NO_2, which makes them potential sensor materials for such molecules.

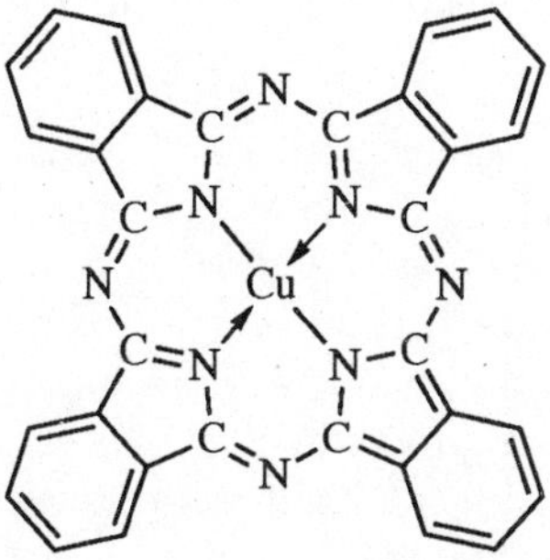

Figure 3.11 Copper phthalocyanine

They have been used in the conductimetric mode in sensors for gases such as NO_2, NO, Cl_2, F_2 and BF_3. The sensitivity and selectivity can be varied by changing the structure of the molecule, in particular the metal, or the substituents on the benzene rings. Unfortunately they often have to be operated at elevated temperatures such as 80–150 °C, and they have slow recovery times, as they form donor–acceptor complexes with the substrate molecules. Lead phthalocyanine will detect down to 1 ppb of NO_2 in air. Electron-donor molecules such as NH_3 and H_2S interfere to some extent, but hydrocarbons do not.

The redox properties of phthalocyanines can be used to provide an electrocatalytic sensor membrane for the voltammetric/amperometric measurement of species such as cysteines, reduced glutathiones, oxalic acid and hydrazines, either in a carbon paste or in a screen-printed carbon electrode.

3.7.7 Screen-printed Electrodes

A recent important development in electrode technology is the screen-printed electrode (Wring and Hart, 1992). The working electrode is usually a graphite powder-based 'ink' printed on to a polyester material. The reference electrode is usually a silver–silver chloride ink. A typical layout is shown in Figure 3.12.

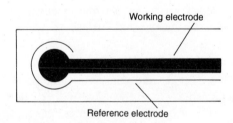

Figure 3.12 A screen printed electrode. Reproduced by permission of the Royal Society of Chemistry from Wang (1994)

Appropriate modifying components can be incorporated into the carbon ink, such as gold, mercury, chelating agents (for use in stripping voltammetry), mediators such as phthalocyanines and ferrocenes to catalyse electron transfer or enzymes such as glucose oxidase, ascorbic acid oxidase, glutathione oxidase or uricase.

The procedure has the advantage of miniaturisation, versatility and cheapness and in particular lends itself to the mass production of disposable electrodes. A version is marketed commercially in the Exactech biosensor for glucose (Figure 3.13).

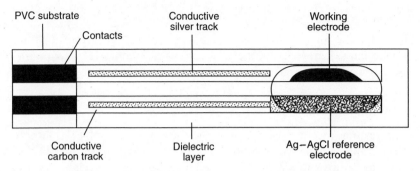

PVC substrate Conductive silver track Working electrode

Contacts

Conductive carbon track Dielectric layer Ag–AgCl reference electrode

Figure 3.13 The ExacTech biosensor disposable electrode strip. Reproduced by permission of the Royal Society of Chemistry from Green and Hilditch (1991)

3.8 EXAMPLES OF APPLICATIONS OF DIFFERENT IMMOBILISATION METHODS

3.8.1 Adsorption

This is rarely used on its own, because of its weak and transitory effect. However, it is commonly used along with other methods. Frequently an enzyme is adsorbed on an electrode surface and then immobilised there with a membrane or other form of encapsulation. In studies of peroxidase-based biosensors, peroxidases could be adsorbed by edge-oriented pyrolytic graphite or glassy carbon. However, it could also be adsorbed on graphite powder prior to the addition of the pasting liquid, especially if the graphite had received some brief heat pretreatment. In this case, there is effectively both adsorption and entrapment in the carbon paste. Experiments showed that liquids other than liquid paraffin gave better results, especially phenylmethylsilicone oil (a component in DC710, a common stationery phase used in gas chromatography). However, even this lacked long-term stability, as the entrapment was really in a mixture, not a polymer matrix. Improvement was made by mixing a carbodiimide with the graphite powder; that one used was 1-cyclohexyl-3-(2-morpholinoethyl)carbodiimide metho-*p*-toluenesulphonate. This facilitated the formation of covalent bonds to hydroxy or carboxyl groups on the surface of the graphite. Addition of the cross-linking agent glutaraldehyde provided further long-term stability.

3.8.2 Microencapsulation (Membrane Entrapment)

This was the method originally used in the first glucose biosensor. A polyethylene membrane, permeable to oxygen, covered the platinum electrode. The glucose oxidase was then sandwiched between this membrane and a cellulose acetate membrane, which is permeable to both oxygen and glucose.

Similarly, the first potentiometric biosensor was constructed from an ammonia ion-selective electrode covered with nylon net to support the urease enzyme, which was held in place with a dialysis membrane.

3.8.3 Entrapment

An interesting recent entrapment procedure used a sol–gel method. A solution of the enzyme was mixed with tetramethoxyorthosilicate (TMOS) at room temperature and pH > 7, followed by gelation and drying. This gave a transparent xerogel with good chemical and thermal stability. It was tested with glucose oxidase for an optical glucose sensor (Braun *et al.*, 1992).

Entrapment is probably the best method of immobilising microorganisms. In a recently reported example from Russia (Reshetilov *et al.*, 1992), *Gluconobacter oxydans* cells were immobilised in polyvinyl caprolactam gel on the gate of a pH-sensitive field-effect transistor (FET). It was tested as a glucose sensor which showed no response to a range of important interfering compounds such as mannitol, sorbose, fructose and lactose.

Entrapment can be combined with electrode surface modification. Bartlett and Caruana (1992) reported the immobilisation of glucose oxidase from a solution containing the enzyme, phosphate buffer (pH 7.0) and phenol. Electrolysis at +0.9 V for 8 min effected polymerisation of the phenol on the electrode surface, thus entrapping the glucose oxidase. They immobilised D-amino acid oxidase in the same way.

3.8.4 Cross-linking

An interesting application of cross-linking was to immobilise urease and penicillinase, as very fine films, directly on the sensitive ends of a glass pH electrode (Meier *et al.*, 1992).

A tyrosinase biosensor for polyphenols was made by pretreating the electrode (gold placed on a ceramic chip) by polymersing pyrrole in 0.1 M tetraethylammonium sulphonate on the surface. The tyrosinase solution was then repetitively coated on the surface alternating with glutaraldehyde solution. This cross-linked the enzyme to the polypyrrole surface (McArdle and Persaud, 1993).

3.8.5 Covalent Bonding

Papain was immobilised on alumina supports as follows: alumina supports were derivatised with organic phosphate links to create free carboxyl groups using a two-step process. Papain was bound to these derivatised aluminas by reaction with the water-soluble 1-ethyl-3-(dimethylaminopropyl)carbodi-imide (Hyndman *et al.*, 1992).

One important aim of attaching enzymes to electrodes is to achieve good electrical contact to facilitate rapid electon transfer. This is particularly important in developing 'third-generation' biosensors which involve direct electron transfer, not via a mediator. Ye *et al.* (1993) described a high current density 'wired' quinoprotein glucose dehydrogenase electrode. The wiring was done with an osmium-containing polyvinylpyridine. A polished glassy carbon electrode was coated sequentially with the polyvinylpyridine complexed with osmium bis(bipyridine) chloride, then glucose dehydrogenase–PQQ (pyrroloquinoline quinone) in HEPES buffer, followed by poly(ethylene glycol 400 diglycidyl ether). After mixing the solutions, the assembly was dried and cured at room temperature for 24 h. An extraordinarily high current of 1.8 mA cm^{-2} was obtained in a buffered glucose solution, although with continuous operation the activity was lost in 8 h.

3.9 REFERENCES

A. M. Abeysekera, J. Grimshaw and J. Trocha-Grimshaw (1992) 'Electroactive polyamino acids, Part 5', *J. Chem. Soc. Perkin 2*, 43.

R. N. Adams (1969) *Electrochemistry at Solid Electrodes*, Marcel Dekker, New York.

S. A. Barker (1987) 'Immobilisation of the biological component of biosensors', in A. P. F. Turner, I. Karube and G. S. Wilson (Eds), *Biosensors: Fundamental and Applications*, Oxford University Press, Oxford, Chap. 6, pp. 85–99.

P. N. Bartlett and D. J. Caruana (1992) 'Electrochemical immobilisation of enzymes. Part V. Microelectrodes for the detection of glucose oxidase immobilised in a poly(phenol) film', *Analyst*, **117**, 1287.

S. Braun, S. Shtelzer, S. Rappaport, D. Avnir and M. Ottolenghi (1992) 'Biocatalysis by sol–gel entrapped enzymes', *J. Non-Cryst. Solids*, **147–148**, 739.

B. R. Eggins (in press) 'Electrochemical sensors and biosensors', in M. R. Smyth, M. E. G. Lyons and V. Cunnane (Eds), *Electrochemical Principles and Applications*. Ellis Harwood, Chichester.

M. W. Espenscheid, A. R. Ghatak-Roy, R. B. Moore, III, R. M. Penner, M. N. Szentirmay and C. R. Martin (1980) 'Sensors from polymer modified electrodes', *J. Chem. Soc. Faraday Trans 1*, **82**, 1051.

M. J. Green and P. I. Hilditch (1991) *Analyst*, **116**, 1217.

J. Grimshaw and S. D. Perera (1990) 'Electrochemical behaviour of poly(thiophene-benzoquinone) films', *J. Electroanal. Chem.*, **278**, 287.

J. Grimshaw and J. Trocha-Grimshaw (1990) 'Electron transfer in a novel synthetic membrane analogue for cytochrome-c', *J. Chem. Soc. Chem. Commun.*, 157.

E. A. H. Hall (1990) *Biosensors*, Open University Press, Milton Keynes, pp. 23–29.

D. Hyndman, R. Burrell, G. Lever and T. G. Flynn (1992) 'Protein immobilisation to alumina supports. II. Papain immobilisation to alumina via organophosphate linkers', *Biotechnol. Bioeng.*, **40**, 1328.

M. E. G. Lyons (1991) 'Electrochemistry—developing technologies and applications', *Annu. Rep. Prog. Chem. Sect. C*, **88**, 135.

M. E. G. Lyons, C. H. Lyons, C. Fitzgerald and T. Banon (1993) *Analyst*, **118**, 361.

F. A. McArdle and C. K. Persaud (1993) 'Development of an enzyme-based biosensor for atrazine detection', *Analyst*, **118**, 419.

H. Meier, F. Lantreibecq and C. M. Tran (1992) 'Application and automation of flow injection analysis (FIA) using fast responding enzyme glass electrodes to detect penicillin in fermentation broth and urea in human serum', *J. Autom. Chem.*, **14**, 137.

A. N. Reshetilov, M. V. Donova and K. A. Koshcheenko (1992) 'Immobilised cells of *Gluconobacter oxydans* as a receptor element of a glucose sensor', *Prikl, Biokhim. Mikrobiol.*, **28**, 518.

A. W. Snow and W. R. Barger (1989) 'Phthalocyanine films in chemical sensors', in C. C. Leznoff and A. B. P. Lever (Eds), *Phthalocyanine Properties and Applications*, VCH, New York.

P. R. Teasdale and G. G. Wallace (1993) 'Molecular recognition using conducting polymers: basis of electrochemical sensing technology', *Analyst*, **118**, 329.

J. Wang (1994) 'Decentralised electrochemical monitoring of trace metals: from disposable strips to remote electrodes', *Analyst*, **119**, 763.

J. Wang and M. S. Lin (1988) 'Mixed plant tissue–carbon paste bioelectrode', *Anal. Chem.*, **60**, 1545.

S. A. Wring and J. P. Hart (1992) 'Chemically modified screen printed electrodes', *Analyst*, **117**, 1299.

L. Ye, M. Hammerle, A. J. J. Olsthoorn, W. Schuhmann, H. -L. Schmidt, J. A. Duine and A. Heller (1993) 'High current density "wired" quinoprotein glucose dehydrogenase electrode', *Anal. Chem.*, **65**, 238.

Chapter 4

Transducers I—
Electrochemistry

There are three basic electrochemical processes which are useful in transducers for biosensors: (i) *potentiometry*, the measurement of a cell potential at zero current; (ii) *voltammetry* (amperometry), in which an oxidising (or reducing) potential is applied between the cell electrodes and the cell current is measured; and (iii) *conductimetry*, where the conductance (reciprocal of resistance) of the cell is measured by an alternating current bridge method.

4.1 POTENTIOMETRY

4.1.1 Cells and Electrodes

If a piece of metal is placed in an electrolyte solution as in Figure 4.1, there is charge separation between the metal (electrode) and the solution. This sets up what we can call an *electron pressure*, usually called a *potential*. It cannot be measured directly, and requires a combination of two such electrode–electrolyte solution combinations. Each is called a *half-cell*. The combination shown in Figure 4.2 is an electrochemical *cell*.

The two half cells must be connected internally by means of an electrically conducting bridge or membrane. Then the two electrodes are connected externally by a potential measuring device, such as a digital voltmeter (DVM). This has a very high internal impedance ($\sim 10^{12}\,\Omega$), such that very little current will flow through it. [If the voltage to be measured is 1 V, then by Ohm's law ($V = IR$), the current $I = 10^{-12}\,A$ (1 pA).]

The electrical circuit is now complete and the e.m.f. of the cell can be measured. This value is the difference between the electrode potentials of the

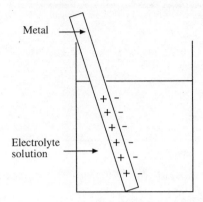

Figure 4.1   A metal electrode dipped in electrolyte solution. One-half cell

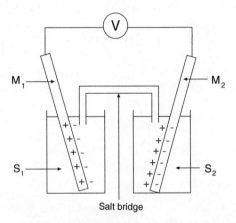

Figure 4.2   Two half-cell electrodes combined, making a complete cell

two half-cells. Its magnitude depends on a number of factors: (i) the nature of the electrodes M_1 and M_2; (ii) the nature and concentrations of the solutions S_1 and S_2; and (iii) the liquid juction potential at the membrane (or salt bridge). The most straightforward situation is if S_1 contains an ion M_1^{n+} of M_1 and S_2 contains an ion M_2^{m+} of M_2.

The practical Daniell cell is a good illustration of such a cell. It is shown in Figure 4.3. It involves copper and zinc electrodes in solutions of copper (II) and zinc(II) sulphates with a porous pot for the bridge. To keep matters simple we shall assume that the concentrations of electrolytes are both 1M. This cell used to be used as a practical battery and has an e.m.f. of 1.10 V.

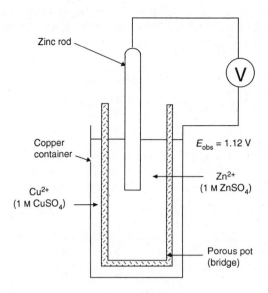

Figure 4.3 The Daniell cell

If we consider each half-cell, we can write the half-cell reactions:

$$Cu^{2+} + 2e^- = Cu \qquad \text{half cell electrode reaction} \qquad (4.1)$$

$$Zn^{2+} + 2e^- = Zn \qquad \text{half cell electrode reaction} \qquad (4.2)$$

If we substract equation 4.2 from equation 4.1 we obtain

$$Cu^{2+} + Zn = Cu + Zn^{2+} \qquad \text{complete cell reaction}$$

The Gibbs free energy for this reaction is negative, showing that the reaction will proceed spontaneously in the direction indicated. The reaction can easily be carried out directly in a test-tube, by the addition of copper(II) sulphate solution to pieces of zinc. The white zinc metal quickly becomes covered with a dark brown coating of copper metal and the blue colour of the copper(II) sulphate solution fades as it is replaced by colourless zinc sulphate. The Gibbs free energy is simply related to the e.m.f. of the cell;

$$\Delta G = -nFE$$

where n is the number of electrons transfered (in this case $n = 2$), F is the Faraday constant $= 96\,487\,C\,mol^{-1}$ and E is the e.m.f. of the cell (we assume that the liquid junction potential is zero). Thus, if ΔG is negative, E is positive.

We may now ask what the ΔG values are for reactions 4.1 and 4.2 separately. If we could find ΔG_1 and ΔG_2, we could find E_1 and E_2 separately.

A simple separation is not possible, however, so we take another approach. Consider the first element of the Periodic Table, **hydrogen**. It is not a metal but it can be oxidised to H^+ by the removal of an electron:

$$H - e^- = H^+$$

which is more usually written as

$$H^+ + e^- = \tfrac{1}{2} H_2$$

ΔG for this reaction is defined as zero for the standard state, the standard state being with $[H^+] = 1\,M$, partial pressure of $H_2 = 1\,atm$ and temperature = 298 K (25 °C). For any standard state the Gibbs free energy is designated $\Delta G^{\ominus}$. The standard electrode potential for hydrogen is therefore

$$E^{\ominus}_{H^+/H_2} = 0$$

We can set up a practical half-cell **hydrogen** electrode. This can then be combined with any other half-cell electrode as in Figure 4.4. For example,

$$Cu^{2+} + 2e^- = Cu \qquad E_1 \tag{4.3}$$

$$2H^+ + 2e^- = H_2 \qquad E_H\,(=0) \tag{4.4}$$

Subtracting equation 4.4 from equation 4.3:

$$Cu^{2+} + H_2 = Cu + 2H^+$$

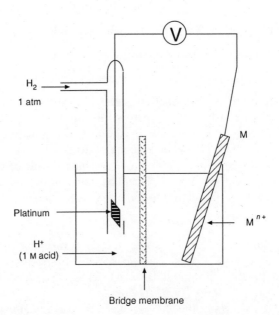

Figure 4.4   A hydrogen electrode connected with another half-cell

Thus

$$E_{cell} = E_1 - E_H = +0.34\,\text{V}$$

Therefore,

$$E_{Cu}^{\ominus} = +0.34\,\text{V}$$

Hence a scale of E values for a whole range of half-cell electrodes can be established **against the standard hydrogen electrode** (SHE), sometimes called the normal hydrogen electrode (NHE).

For the other half of the Daniell Cell, the zinc electrode,

$$Zn^{2+} + 2e^- = Zn \qquad E_{Zn} \qquad\qquad (4.5)$$

$$2H^+ + 2e^- = H_2 \qquad E_H\,(=0) \qquad\qquad (4.6)$$

Subtracting equation 4.6 from equation 4.5:

$$Zn^{2+} + H_2 = Zn + 2H^+$$

Thus

$$E_{cell} = E_{Zn} - E_H$$

Therefore

$$E_{Zn}^{\ominus} = -0.76\,\text{V}$$

Combining the half-cell e.m.f.s for copper and zinc gives the cell e.m.f. for the Daniell cell

$$E_{cell} = +0.34 - (-0.76) = 1.10\,\text{V}$$

4.1.2   Reference Electrodes

The standard hydrogen electrode is a *reference electrode* (RE), i.e. one to which other electrodes may be referred. While it is not difficult to set up an SHE in the laboratory, it is not very convenient for routine measurements as it involves flowing hydrogen gas, which is potentially explosive. Other secondary reference electrodes are used in practice which are easy to set up, are non-polarisable and give reproducible electrode potentials which have low coefficients of variation with temperature.

Many varieties have been devised but two are in common use and are easy to set up and are available commercially.

(*i*)   *The silver–silver chloride electrode.* Silver chloride has the advantage of being sparingly soluble in water. The half-cell reaction is

$$AgCl + e^- = Ag + Cl^- \qquad E^{\ominus} = +0.22\,\text{V}$$

The electrode consists of a silver wire coated with silver chloride dipping into a solution of potassium chloride (usually 1 M). The silver chloride coating can easily be obtained by making the silver wire the anode in an electrochemical cell with a platinum cathode and a potassium chloride electrolyte and electrolysing for about 30 min with a positive potential of about 0.5 V applied to the silver. The surface of the silver metal is oxidised to silver ions, which attract chloride ions to form a surface layer of silver chloride. This is shown in Figure 4.5.

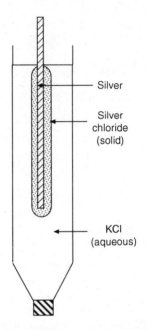

Figure 4.5   A silver/silver chloride reference electrode

(ii)   The saturated calomel electrode (SCE). **Calomel** is the old-fashioned name for mercury(I) chloride (Hg_2Cl_2), which, like silver chloride, is sparingly soluble in water. The half-cell reaction is

$$Hg_2Cl_2 + 2e^- = 2Hg + 2Cl^- \qquad E^{\ominus} = +0.24\,V$$

It consists of a mercury pool in contact with a paste made by mixing mercury, mercury(I) chloride powder and saturated potassium chloride solution, the whole being in contact with a saturated solution of potassium chloride. A saturated solution can be easily made by shaking potassium chloride with water until no more dissolves. This gives a constant, reproducible concentration without the need for weighing and measuring. The electrode is shown in Figure 4.6.

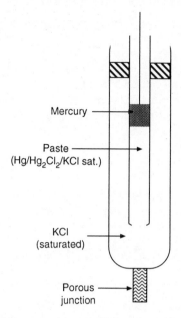

Figure 4.6   A saturated calomel electrode (SCE)

These electrodes are suitable for most purposes in aqueous solutions. Other types are available for use in non-aqueous solutions or if chloride ions must be absent.

One usually measures the potential difference between an indicator electrode and the reference electrode to give the cell e.m.f.:

$$E_{cell} = \Delta E = E_{ind} - E_{RE}$$

4.1.3   Quantitative Relationships. The Nernst Equation

So far we have only considered electrode potentials at one concentration of Ox or R, usually 1 M. Now we must consider the effect of different concentrations on the electrode potential. This is of fundamental importance for analytical applications of potentiometry. The basic Nernst equation is a logarithmic relationship derived from fundamental thermodynamic equations such as

$$\Delta G = -RT \ln K$$

So, for the half-cell reaction, $Ox + ne^- = R$, the Nernst equation is

$$E = E^\ominus + \frac{RT}{nF} \ln \left(\frac{a_{Ox}}{a_R} \right)$$

where a_{Ox} and a_R are activities, i.e. ideal thermodynamic concentrations, which for dilute solutions can be taken to be the same as concentrations. It is usually more useful to express concentrations in powers of ten and therefore to use logarithms to base 10 rather than the *natural* logarithms to base **e**. The Nernst equation then becomes

$$E = E^\ominus + 2.303 \frac{RT}{nF} \log \left(\frac{[Ox]}{[R]} \right)$$

It should be noted that this equation has the same form as the Henderson–Hasselbach equation for the pH of mixtures of acids and bases:

$$pH = pK_a + \log \left(\frac{[A^-]}{[HA]} \right)$$

Logarithms are less familiar to students now that 'log tables' are no longer an essential tool for calculations—multiplications, divisions, powers and roots. However, the idea of the logarithm is still essential for much scientific work. It is a way of expressing a value in terms of a power. Thus, we commonly say that $2 \times 2 = 2^2 = $ 'two squared' = 4. Similarly, $2 \times 2 \times 2 = 2^3 = $ 'two cubed' = 8, and so on. The power to which 2 is raised, i.e. 2 or 3, is called the 'logarithm' (to base 2) of the numbers 4 and 8, respectively. When we measure very large numbers such as the speed of light, which is $300\,000\,000$ m s^{-1}, it is often more convenient to express them as powers of ten, thus 3×10^8 m s^{-1}. Similarly, very small numbers such as concentrations of very dilute solutions, e.g. $0.000\,000\,01$ M, i.e. $1/100\,000\,000$ M, can better be expressed as 1×10^{-8} M. In these examples the power or indices $+8$ and -8 are the logarithms (to base 10). In scientific work it happens that a special base for logarithms called 'e' is used. These are called 'natural logarithms'. The value of e is $2.718.\ldots$ This need not worry us as logarithms to base e are simply related to logarithms to base 10 by the value 2.303. Thus the natural logarithm of x (called $\ln x$ or $\log_e x$) is just $2.303 \log_{10} x$. In fact, with a calculator this conversion is rarely needed as scientific calculators give values for both natural logarithms (to base e) and common logarithms (to base 10). To get back to a number from a logarithm, one needs the index or anti-logarithms as it used to be called. This is also given by the calculator.

A calculator key is labelled

10^x
LOG

The main function gives the common logarithm (to base 10) of a number. The inverse (or second) function gives the anti-logarithm 10^x.

The adjacent key will be labelled

e^x
LN

This similarly gives the natural logarithm (1n) of x and the inverse gives e^x. A little practice is needed with these concepts if they are unfamiliar, so that facility may be acquired with the manipulation of concentrations in terms of powers of ten.

Measurements of acidity (hydrogen ion concentrations) are normally expressed as pH values, where pH *is* a logarithm. Thus, for an acid concentration of 10^{-3} M we take the positive value of the negative index, i.e. 3, so the pH value is 3. In general,

$$pH(x) = -\log_{10}(x)$$

A slightly more complex, but very, common example relates to the pH of 1 M acetic acid (a weak acid). The hydrogen ion concentration is 1.8×10^{-5}, so pH $= -\log_{10}$ $(1.8 \times 10^{-5}) = 4.745$. Thus $10^{-4.745} = \log (1.8 \times 10^{-5})$.

If we put in the values of the constants R and F and a value for T, we obtain

$$\frac{RT}{F} = 0.0257 \, V \qquad \text{at } T = 298 \, K \, (25 \, ^\circ C)$$

$$\frac{RT}{F} = 0.0252 \, V \qquad \text{at } T = 293 \, K \, (20 \, ^\circ C)$$

$$2.303 \, \frac{RT}{F} = 0.0591 \qquad \text{at } T = 298 \, K$$

$$2.303 \, \frac{RT}{F} = 0.0580 \qquad \text{at } T = 293 \, K$$

It is often useful to approximate these latter values to 0.06, giving a simplified form of the Nernst equation:

$$E = E^\ominus + 0.06 \log \left(\frac{[Ox]}{[R]} \right)$$

The reduced species, R, is often a metal, in which case it has a constant concentration (activity) = 1, so the equation simplifies further to

$$E = E^\ominus + 0.06 \log[Ox]$$

We can generalise this for practical situtations in which $E^\ominus$ and 2.303 RT/F may not be known or may differ from the theoretical values:

$$E = K + S \log[Ox]$$

This equation is a very useful practical form of the Nernst equation. As we shall see from experimental data, we can plot a graph of E against log [Ox] which would normally give a straight line of slope S and intercept K. Then the experimental values of S and K can be compared with the theoretical values, $S = 2.303 \, RT/F$ and $K = E^\ominus$.

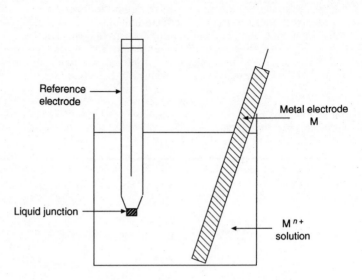

Figure 4.7   A reference electrode combined with another half-cell

If we now incorporate the reference electrode potential (E_{REF}) and the liquid junction potential ($E_{1,j}$), as in Figure 4.7:

$$E_{cell} = E'_{M,M}n+ - E_{RE} - E_{1,j}$$

$$E'_{M,M}n+ = E'^{\ominus} + S\log[M^{n+}]$$

Therefore,

$$E_{cell} = (E'^{\ominus} - E_{RE} - E_{1,j}) + S\log[M^{n+}]$$

Hence

$$E_{cell} = K + S\log[M^{n+}]$$

with

$$K = E'^{\ominus} - E_{RE} - E_{1,j}$$

4.1.4   Concentration Cells

If instead of a reference electrode we incorporate a similar half-cell with the same redox couple but with a different concentration of Ox, as in Figure 4.8, we can set up the two half-cell reactions as follows:

$$E_1 = E^{\ominus} + S\log[Ox]_1$$

$$E_2 = E^{\ominus} + S\log[Ox]_2$$

Subtracting:

$$\Delta E = E_1 - E_2 = S \log\left(\frac{[Ox]_1}{[Ox]_2}\right)$$

Now, if $[Ox]_2$ is kept constant (perhaps at a reference concentration), we obtain

$$\Delta E = \text{constant} + S \log[Ox]_1$$

where the constant is $-S \log[Ox]_2$.

This result is made use of in a practical way with most forms of ion selective electrode. The general practical arrangement is shown in Figure 4.8.

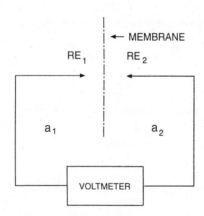

Figure 4.8 A concentration cell. RE_1 and RE_2 = reference electrodes

On the left is the test solution to be determined into which is dipped a reference electrode. The ion-selective membrane is in the middle dividing the test solution from the standard solution on the right, which consists of a fixed concentration of the ions being measured. Into this is placed a second reference electrode. The two reference electrodes are connected through the high-impedance voltmeter—usually a digital voltmeter. The electrode system is then usually calibrated with standard solutions in one of a number of ways, as described below.

The observed voltage is the difference between the two half-cell electrodes, which are identical, except that the concentrations of the ion being determined differ in each half of the cell. We can write the voltage (e.m.f.) of the cell as follows:

$$E = E_{RE1} + E_{RE2} + E_{1j} - S \log a_2 + S \log a_1$$

So

$$E = K + S \log a_1$$

where

$$K = (E_{RE2} - E_{RE2} + E_{1j} - S \log a_2)$$

a_1 = activity of test solution, and a_2 = activity of reference (standard) solution.

4.1.5 Practical Aspects of Ion-selective Electrodes

In order to obtain consistent, reproducible results with the lowest detection limits, certain precautions have to be observed. Sample standardisation does not involve extensive pretreatment. Usually the addition of a special buffer is sufficient. The following factors may need to be observed.

(i)   The ionic strength needs to be kept constant from one sample to the next. This can simply be done by adding a fairly high, constant concentration of an indifferent electrolyte, i.e. one that does not interfere in any way, to each sample and each standard.

(ii) The pH may need to be controlled at a certain level. This is more important with some ionic samples than others, e.g. fluoride.

(iii) It may be possible and desirable to add components which minimise or eliminate interfering ions.

Appropriate mixtures to provide these properties are usually called ISAs—ionic strength adjusters—or more fully TISABs—total ionic strength adjustment buffers. The best way to illustrate these is by two examples. For nitrate ISEs the ISA is commonly just sodium sulphate. This, although not strictly a pH buffer, keeps the pH well within the required 2–12 limits. However, it does not eliminate the considerable and important interferences from chloride and nitrite. Alternative ISAs of more complex composition will do so. Silver sulphate was used to precipitate out the chloride, but has now been replaced by a more complex but less expensive lead acetate mixture. Fluoride electrodes require a much more complex mixture made up as follows: 1 M sodium chloride (to control the ionic strength) + 1 M acetic acid, adjusted to pH 5.5 with sodium hydroxide (to control the pH, as a high pH would result in interference from hydroxide) + 10^{-3} M sodium citrate. Citrate is a good complexing agent for Al^{3+} and Fe^{3+}, both of which interfere by themselves complexing with fluoride.

Temperature needs to be controlled for accurate work, as the Nernst factor includes temperature and a 1 K variation in temperature will cause a 2% error in measured concentration. Similarly, other experimental factors such as stirring and equilibration time (usually 30 s–2 min) before reading must be standardised.

4.1.6 Measurement and Calibration

4.1.6.1 CALIBRATION GRAPH AND DIRECT READING

This is the most straightforward method. A series of standard solutions are made up with added ISA and the potentials are measured. Then a calibration graph is plotted of voltage against log(concentration). Deviations from linearity or Nernstian slope do not matter. The sample is treated in the same way and its log(concentration) value is read from the graph. Fresh calibration graphs should be prepared regularly.

4.1.6.2 STANDARD ADDITION

The sample is prepared as before and its voltage read. Then a known amount of a standard of higher concentration, usually about ten times the expected sample concentration, is added and a second voltage reading is taken. The data are then fitted to an equation which should include a correction for dilution by the added standard.

If C_u = unknown concentration in V_u cm^3 of solution and C_s = added standard concentration in V_s cm^3 of solution, then

$$E_1 = K + S \log C_u$$

$$E_2 = K + S \log(C_u V_u + C_s V_s)/(V_u + V_s)$$

Subtracting:

$$E = S \log\{Cu/[C_u V_u + C_s V_s)/(V_u + V_s)]\}$$

This can be rearranged to give

$$C_u = C_s/\{10^{E/S}[1 + (V_u/V_s)] - V_u/V_s\}$$

hence C_u can be obtained.

4.1.6.3 GRAN PLOT

This is really an extension of the standard addition method using multiple standard additions. The procedure is the same as in the standard addition method except that several additions are made (say five or more). Using the above nomenclature, except that the single value C_s is replaced by a variable C_s, where C_s represents the increase in concentration in the sample solution produced by each addition, we have

$$E = K + S \log(C_u + C_s)$$

$$E/S = K/S + \log(C_u + C_s)$$

Take antilogs:

$$10^{E/S} = K'(C_u + C_s)$$

where $K' = 10^{K/S}$. $10^{E/S}$ is plotted against C_s as shown in Figure 4.9. The graph is a straight line with a negative intercept $-C_u$, as when $10^{E/S} = 0$, $C_u = -C_s$. This derivation does not show corrections for added volumes of standards.

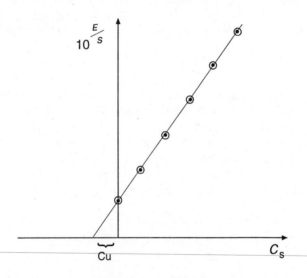

Figure 4.9 Grans' plot

4.1.7 Examples of Ion-selective Electrodes

4.1.7.1 GLASS MEMBRANE TYPE

The best, indeed almost universally known, example is the glass membrane electrode for measuring hydrogen ion concentration or acidity, usually called the pH electrode. The thin glass membrane is highly selective to hydrogen ions over a very wide range of concentrations. The composition of the glass is critical for this performance. If it is changed it may make the glass membrane selective to other ions. The usual composition for hydrogen ions is 22% Na_2O : 6% CaO : 72% SiO_2.

A typical pH combination electrode is shown in Figure 4.10. This type incorporates the second reference electrode in a concentric glass tube round the main electrode tube. Contact between this electrode and the test solution is through a small glass frit. The two reference electrodes are normally of the Ag/AgCl type. The hydrogen ion glass electrode is usually expressed as a 'pH' electrode and calibrated in terms of pH rather than hydrogen ion activity, where

$$pH = -\log a_{H^+}.$$

So

$$E_{obs} = K + 0.0591 \log a_{H^+} = K - pH$$

and therefore

$$pH = \frac{K - E}{0.0591}$$

Other glass membrane ion-selective electrodes can be obtained for Na^+, Li^+, K^+ and Ag^+.

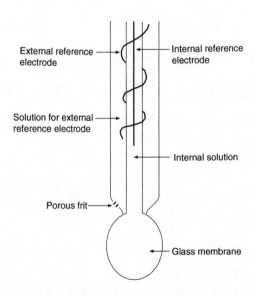

Figure 4.10 A pH electrode

4.1.7.2 SOLID-STATE TYPE

The simple diagram in Figure 4.11 exemplifies the general structure for this type of electrode. It normally has a separate reference electrode provided by the worker to dip into the test solution. It may be, but need not be, the same as the internal reference electrode built in by the manufacturer. The solid-state membrane can be a solid crystal such as LaF_3 in the fluoride electrode or a pressed pellet of powdered material such as Ag_2S in sulphide electrodes.

Some examples of this type are F^-, Cl^-, Br^-, I^-, SCN^- and S^{2-}. Of these, the fluoride electrode is regularly used in water treatment plants for measuring the fluoride levels in drinking water.

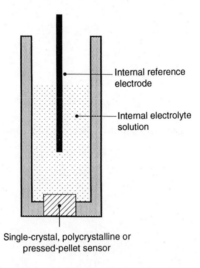

Internal reference
electrode

Internal electrolyte
solution

Single-crystal, polycrystalline or
pressed-pellet sensor

Figure 4.11 A solid-state ion-selective electrode

4.1.7.3 LIQUID ION-EXCHANGE MEMBRANE TYPE

The membrane is made of a hydrophobic material such as plasticised PVC. Absorbed into this membrane is the liquid ion-exchange material such as valinomycin (for potassium). In order to maintain the concentration level in the membrane there is a reservoir of the ion-exchange liquid dissolved in an organic solvent. Figure 4.12 shows the details of this type including the special reservoir for the ion exchanger solution and also the reference solution and internal reference electrode.

Some examples of this type are NO_3^-, Cu^{2+}, Cl^-, BF_4^-, ClO_4^- and K^+. The nitrate electrode is used extensively for the measurement of nitrate in soils and waters.

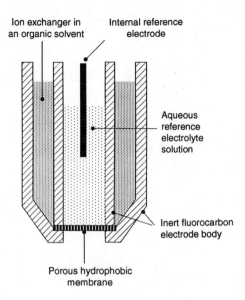

Figure 4.12 A liquid ion-exchange membrane ion-selective electrode

4.1.7.4 GAS-SENSING ELECTRODES

These are mainly based on pH electrodes and can detect gases which in aqueous solution form acidic or basic solutions. A gas-permeable membrane is attached as shown in Figure 4.13. Between the membrane and the hydrogen-selective glass membrane is an internal electrolyte containing material which will form a buffer with the gas material. For example, for the ammonia electrode ammonium chloride is used, so that an equilibrium is set up thus:

$$NH_4Cl = NH_4^+ + Cl^-$$

$$NH_3 + H^+ = NH_4^+$$

$$K_a = \frac{[NH_3][H^+]}{[NH_4^+]}$$

Then

$$\log[NH_3] = pH + pK_a + \log[NH_4^+]$$

The presence of the high concentration of ammonium chloride keeps the concentration of ammonium ions constant. Hence the logarithm of the ammonia concentration is directly porportional to the pH of the solution.

Electrodes for SO_2, NO_2 and H_2S are made in a similar way.

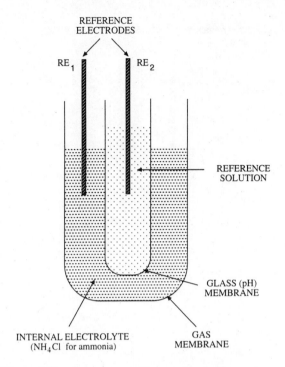

REFERENCE
ELECTRODES

RE$_1$ RE$_2$

REFERENCE
SOLUTION

GLASS (pH)
MEMBRANE

INTERNAL ELECTROLYTE   GAS
(NH$_4$Cl for ammonia)   MEMBRANE

Figure 4.13   A gas-permeable membrane electrode

4.1.8   Use of Ion-selective Electrodes in Biosensors

Relatively few ISEs are used in biosensors. Those most used are H$^+$, NH$_4^+$
and NH$_3$, which are based on the pH principle. Occasionally a CO$_2$, an I$^-$ or
perhaps an S^{2-} electrode may be used. Examples are given in more detail in
Chapter 8.

4.2   VOLTAMMETRY (AMPEROMETRY)

4.2.1   Linear Sweep Voltammetry

The above terms cover a range of techniques involving the application of a
linearly varying potential between a working electrode and a reference
electrode in an electrochemical cell containing a high concentration of an
indifferent electrolyte to make the solution conduct—called the *supporting
electrolyte*—and an oxidisable or reducible species—the *electroactive species*.
The current through the cell is monitored continuously. A graph is traced on

a recorder of current against potential, which is called a *voltammogram*. The straightforward technique is called *linear sweep voltammetry*. A typical voltammogram is shown in Figure 4.14.

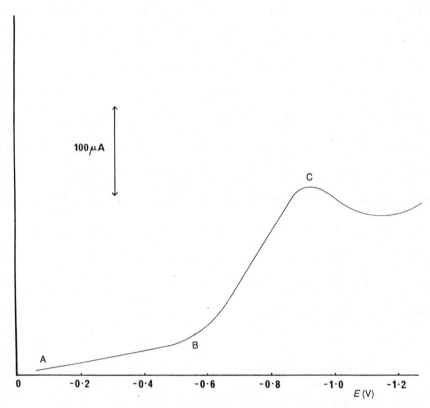

Figure 4.14 A linear sweep voltammogram

At the start (point A) the current is very small. Between points A and B it rises very slowly owing to the residual current (from impurities) and double-layer charging (the electrode–solution interface acting as a capacitor). Sometimes this is called the background current. At point B the potential approaches the reduction potential of the Ox. The increasing potential causes electrons to transfer from the electrode to Ox at an increasing rate according to the general reaction

$$Ox + ne^- \longrightarrow R$$

The increasing rate of reduction causes the cell current to increase. It can be shown that the net cell current in this region is given by the algebraic sum of a cathodic (reduction) current (i_c) and an anodic current (i_a). Hence

$$i_{net} = i_c + i_a$$

where

$$i_c = nFAk^0_f C_{Ox} \exp[-\alpha nF(E - E_{eq})/RT]$$

$$i_a = -nFAk^0_b C_R \exp[(1 - \alpha)nF(E - E_{eq})/RT]$$

As E (the applied potential) increases, i_c increases and i_a decreases, causing the observed rise in the voltammetric wave. This rise does not continue indefinitely, however, as it is limited by the fact that the concentration of Ox becomes depleted by the reduction process and the current is limited by the decreasing rate of diffusion of fresh Ox from the bulk of the solution. This results in a peak in the current at point C in Figure 4.14. This diffusion effect is shown qualitatively in Figure 4.15.

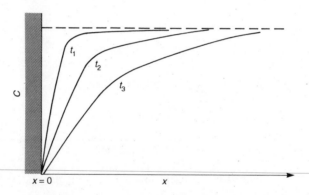

Figure 4.15   The diffusion effect at an electrode for a potential step redox process

The value of the diffusion-limited current is obtained from Fick's first law of diffusion:

$$\text{flux of material to electrode surface} = \frac{dN}{dt} = D\frac{dC}{dx}$$

$$i_d = nFAD\frac{dC}{dx}$$

The progressive depletion effect with time, shown in Figure 4.15, is expressed by Fick's second law of diffusion:

$$\frac{dC}{dt} = D\frac{d^2C}{dC^2}$$

Obtaining even simple solutions to this equation requires fairly advanced mathematics, and for more complex solutions only approximate solutions are possible using numerical methods of integration, with a computer. However, some of the resulting expressions are fairly straightforward. For our purposes, the most important result is an expression for the current at the peak in Figure 4.14 (point C). For the reversible electron transfer situation the peak current is given by the Randles–Sevcik equation:

$$i_p = 2.68 \times 10^5 \, n^{3/2} A D^{1/2} C_{Ox} \, v^{1/2} \qquad \text{(at 298 K)}$$

If the reverse reaction does not occur or if the electron transfer rates are slow (small values of k_f and k_b), the reaction is called *irreversible* and the equation is modified to become

$$i_p = 2.98 \times 10^5 \, n(\alpha n_a) A D^{1/2} C_{Ox} \, v \qquad \text{(at 298 K)}$$

The important thing to notice is that in both cases *the peak current is directly proportional to the concentration of Ox*. This is the most important result for use in analysis.

The potential at which the peak occurs, E_p, is related to the standard redox potential for the couple Ox/R by the equation

$$E_p = E^O + \frac{0.056}{n}$$

for the reversible situation. For the irreversible situation other factors are also involved.

4.2.2 Cyclic Voltammetry

While the amount of Ox at the electrode surface becomes depleted by the reduction, it is of course replaced by R which diffuses away into the solution. Hence if we reverse the potential sweep from the positive side of the peak we shall observe the reverse effect. As the potential sweeps back towards the redox potential, the R species will start to be reoxidised to Ox. The current will now increase in the negative (oxidising) direction until an oxidation peak is reached. Figure 4.16 shows the potential time scan. The overall *cyclic voltammogram* is shown in Figure 4.17.

Figure 4.17 shows two peaks, one corresponding to the reduction of the

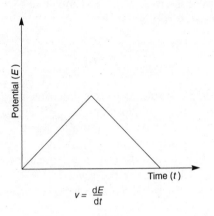

$$v = \frac{\mathrm{d}E}{\mathrm{d}t}$$

Figure 4.16   A potential–time scan

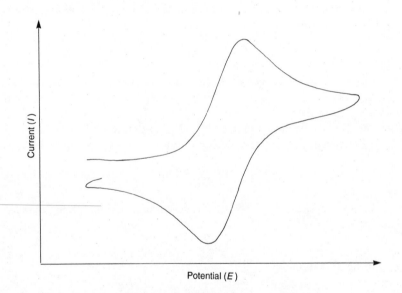

Figure 4.17   A reversible cyclic voltammogram

original substrate and the second corresponding to the reoxidation of the product back to the original substrate. The peak currents are of almost identical heights. The peak potentials are shifted by $0.056/n$ V relative to each other. The average of the two peak potentials is equal to the standard redox potential regardless of the concentration of substrate or its diffusion coefficients or rates of electron transfer.

If the electrode process is irreversible, the peaks will be shifted away from each other, so that $E_{p(c)} - E_{p(a)} > 0.056/n$. However, the average of the two peak potentials is still a good measure of E^O. Sometimes the reverse peak will not be present at all or will have a different or distorted shape. This indicates complete irreversibility due to further reaction of the initially formed reduction product. This is best shown by an example. 4-Acetylaminophenol (paracetamol) is easily oxidised electrochemically at a carbon paste electrode. The cyclic voltammogram is shown in Figure 4.18.

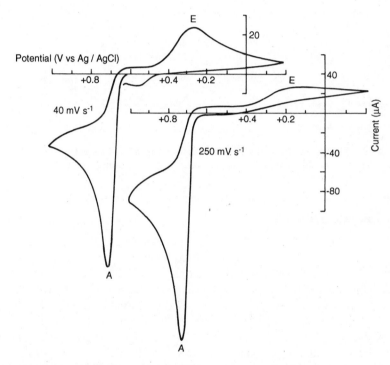

Figure 4.18 Cyclic voltammetry of paracetamol. Reproduced by permission of the Journal of Chemical Education from Benschoten *et al.* (1983)

As the initial reaction is an oxidation (removal of electrons), the initial peak is downwards. On the reverse scan, the corresponding reduction peak is well separated from the initial oxidation peak. It is also much broader. A study of this system at three different sweep rates and three different pH values elucidates the reaction mechanism and explains the behaviour of the cyclic voltammograms. Figure 4.19 shows the mechanism of the reaction.

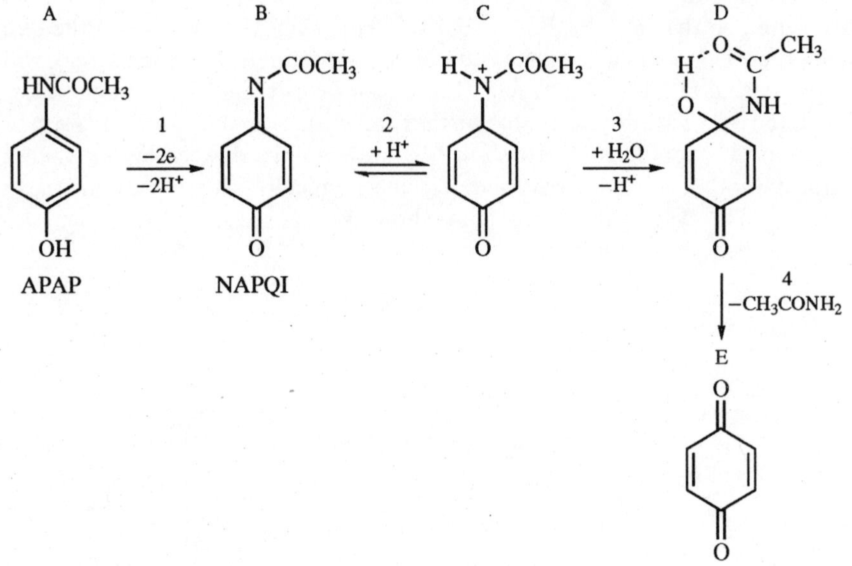

Figure 4.19 The mechanism of voltammetric oxidation of paracetamol. Reproduced by permission of the Journal of Chemical Education from Benschoten *et al.* (1983)

4.2.3 Chronoamperometry

A different, but closely related, technique which clearly shows diffusion control is chronoamperometry. In modified form it is particularly useful with biosensors. Instead of sweeping the potential it is stepped in a square-wave fashion (Figure 4.20) to a potential just past where the peak would be in linear sweep voltammetry. The current is then monitored as a function of time. It decays because of the collapse (or spreading out) of the diffusion layer as shown in Figure 4.15. The simplest solution to the diffusion equation, which can be obtained analytically, shows that the decay is proportional to the reciprocal of the square root of time, as shown in the Cottrell equation:

$$i_d = \frac{nFADC_{Ox}}{\pi^{1/2}t^{1/2}}$$

The current–time profile is shown in Figure 4.20.

Chronoamperometry can be used to determine any of the variables in the equation, knowing the others. Usually it is used to determine n, A or D and sometimes C. The term $i_d/t^{1/2}$ can be determined from the data which form Figure 4.20. The different variables can then be obtained.

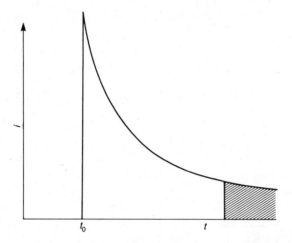

Figure 4.20 The current–time profile associated with a potential step redox process

4.2.4 Amperometry

This is the usual name for the analytical application of the chrono-amperometric technique. With certain cell and electrode configurations, the decaying current reaches an approximately steady state after a certain time. This is shown by the shaded part of the curve in Figure 4.20. The current has become effectively independent of time, as indicated by the equation

$$i = \frac{nFADC_{Ox}}{\delta}$$

where δ is a constant related to the diffusion layer thickness. This relationship is much more useful for analytical work, even though the current has decayed considerably from its highest values.

4.2.5 Kinetic and Catalytic Effects

In most applications involving biosensors, simple reversible or irreversible reactions are rare. Usually there is a coupled chemical reaction, with maybe just a proton transfer, but is sometimes a more complex reaction—as in the example of paracetamol given above in Figures 4.18 and 4.19. A generalised equation is

$$Ox + ne^- = R$$

$$R + A \xrightarrow{k} B$$

Another effect which is particularly useful and common with biosensors is the *catalytic reaction*, in which the original reactant Ox is regenerated in the follow-up reaction. This is shown in the equation

$$R + ne^- = Ox$$

$$Ox + A \xrightarrow{k} R + B$$

The detailed analysis of this situation is complicated but is not needed for understanding the operation of a biosensor.

The effect of this reaction is that the redox process cycles round many times. The reverse reduction of Ox is not seen but the forward oxidation peak is enhanced many times. This is shown in Figure 4.21.

The reversible cyclic voltammogram is of ferrocene (dicyclopentadieneiron(III)). The catalytic wave is caused by interaction with glucose oxidase in the presence of glucose (described in more detail in Chapter 8).

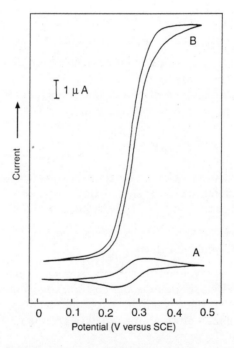

Figure 4.21 A catalytic cyclic voltammogram showing (A) cyclic voltammogram of ferrocene moncarboxylic acid in the presence of glucose and (B) as for (A) but with the addition of glucose oxidase. Reproduced by permission of Analytical Chemistry from Cass *et al.* (1984)

4.3 CONDUCTIVITY

Conductivity is the inverse of resistance. It is a measure of the ease of passage of electric current through a solution. Ohm's law gives the relationship

$$E = IR$$

and conductance, L [in siemens (S); $1S = 1\Omega^{-1}$]:

$$L = 1/R$$

so

$$E = I/L$$

Conductance is related to the dimensions of a cell in a similar way to resistance. For a cell of length l and cross-sectional area A, the conductance $L = \kappa A/l$ where κ is the specific conductivity (S cm^{-1}). It is often further normalised by dividing by the morality of the solute to give the molar conductivity, $\Lambda = \kappa/C$ (C in mol cm^{-3}), so the units of Λ are S mol^{-1} cm^2.

Conductivity is fairly simple to measure. It is directly proportional to the concentration of ions in the solution. Figure 4.22 shows the general conductivity bridge circuit. In the traditional bridge R_2 is adjusted to balance the bridge, and a cell constant used to convert conductance into (specific) conductivity. In modern instruments this is done automatically to give a digital readout.

The conductivity varies according to the charge on the ion, the mobility of the ion and the degree of dissociation of the ion. These introduce

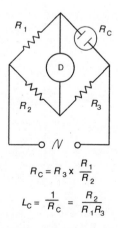

$$R_C = R_3 \times \frac{R_1}{R_2}$$

$$L_C = \frac{1}{R_C} = \frac{R_2}{R_1 R_3}$$

Figure 4.22   A conductivity bridge

complications. Also, in itself the technique has no selectivity. It can be used in controlled situations but really needs to have selectivity superimposed by means of a membrane or coating. Several papers have discussed the application of modified electrodes for the determination of gases using conductivity measurements. Some interesting applications have recently been made using multi-electrode conductors coated with different polypyrroles with different counter ions. In one case an array of five differently coated electrodes was used to distinguish alcoholic beverages (beers or spirits). In another case an array of twelve electrodes was used to distinguish the characteristics of different beers. It was referred to as an *electronic nose*. A statistical method of pattern recognition analysis was used, related to the neural network technique. This could also be biological in nature, leading to another kind of biosensor.

The measurement of conductance involves an alternating current, in the classical conductance bridge. This can be extended by varying the frequency of the alternating current. The quantity measured is then the *admittance* (= 1/*impedance*), which not only depends on simple conductance but also on the capacitance and inductance of the system. These components can be separated as imaginary components, particularly using a frequency response analyser, and displayed in an Argand diagram as in Figure 4.23. These are sometimes called admittance (impedance) spectra. This approach has not yet been used greatly in developing sensors and biosensors, but is receiving increasing attention.

Potentially, a change in conductance can be used to follow any reaction that produces a change in the number of ions, the charge on ions, the

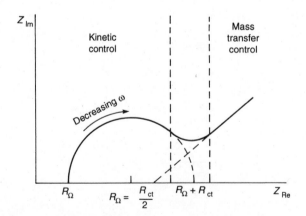

Figure 4.23 Argand diagram showing the frequency dependence of 'imaginary' impedance against the 'real' impedance. Reproduced by permission of J. Wiley & Sons from Bard and Faulkner (1980)

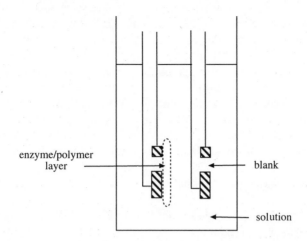

Figure 4.24 A differential mode of conductivity cell used in biosensors

dissociation of ions or the mobility of ions. Usually a differential type of cell is used, as shown in Figure 4.24.

Some examples are as follows:

(i) urea + $2H_2O \xrightarrow{\text{urease}} 2NH_4^+ + HCO_3^-$

This clearly involves a change in ions, and can be followed conductimetrically with improved speed and sensitivity compared with colorimetric methods. Conductance measurements are unaffected by colour or turbidity.

(ii) Many other enzymes result in appropriate changes in conductivity:
amidases: generate ionic groups;
dehydrogenases and decarboxlases: result in separation of unlike charges;
esterases: involve proton migration;
kinases: cause change of degree of association of ions;
phosphatases and sulphatases: result in change in size of charge-carying groups.

4.4 FIELD EFFECT TRANSISTORS

Field effect transistors (FETs) are devices in which a transistor amplifier is adapted to be a miniature transducer for the detection and measurement of potentiometric signals, produced by a potentiometric sensor process on the gate of the FET. A separate reference electrode is needed. Circuit wiring is minimised, so that in addition to miniaturisation, electronic noise is greatly reduced and sensitivity is increased. The FET device can be part of an integrated circuit system leading to the readout, or to the processing of the analytical data. However, no very satisfactory miniaturised reference electrodes exist. According to Janata (1989), most of the proposed versions violate some of the basic principles of reference electrodes. Despite that, a number of possibilities have been proposed and used, varying from a 'pseudo-reference electrode', consisting of a single platinum or silver wire, to the screen-printed type made with silver–silver chloride ink. Perhaps a more satisfactory approach is to avoid the problem by operating in a differential mode with two FETs, one a blank with a gate having negligible response to the analyte and the other coated with the analyte–selective membrane.

The basic type of FET is the insulated gate FET (IGFET). This is shown in Figure 4.25. A source region (4) consisting of n-type silicon is separated from a similar drain (5) region, also of n-type silicon, by p-type silicon (1) and the

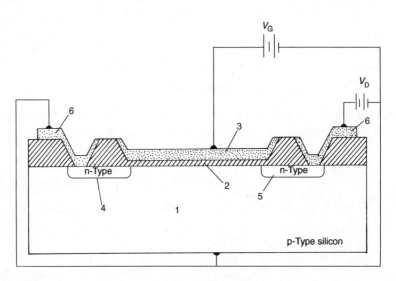

Figure 4.25 Diagram of the IGFET. 1, p-Type silicon substrate; 2, insulator; 3, gate metal; 4, n-type source; 5, n-type drain; 6, metal contacts to source and drain

insulator (2) consisting of silicon dioxide. The source is electrically biased with respect to the drain by the applied potential, V_D. The gate (3) is a metal, insulated from the rest, so that it forms a capacitor sandwich, metal/insulator/semiconductor (MIS), as shown in Figure 4.26.

This gate region is charged with a bias potential V_G. The current from the drain (5) to the source (4), I_D, is measured. There is also a threshold

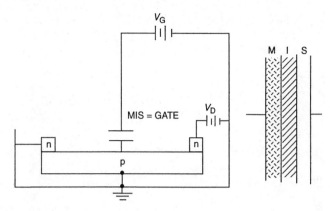

Figure 4.26 The gate in an IGFET. M = metal; I = insulator; S = semiconductor

potential, V_T, at which silicon changes from p-type to n-type, and inversion occurs.

With a small positive V_D and $V_G < V_T$, silicon (1) remains in the p-state, and there is no drain current. n-Si is biased positive with respect to p-Si. When $V_G > V_T$ there is surface inversion, and p-Si becomes n-Si. Now current can pass from drain to source, without crossing the reversed bias p–n junction. V_G now modulates the number of electrons from the inversion layer and controls conductance. I_D flows from source to drain, and is proportional to the electrical resistance of the surface inversion layer and proportional to V_D.

In order to convert this device into a sensor, the metal of the gate is replaced by a chemically sensing surface. This general conformation is known as a CHEMFET and is shown in Figure 4.27.

The chemically sensitive membrane (3) is in contact with the analyte solution (7). A reference electrode (8) completes the circuit via the V_G bias. V_G is corrected by the membrane potential minus the solution potential:

$$\phi_{\text{mem}} - \phi_{\text{sol}} = 1/n_i F(\mu_{i\text{sol}} - \mu_{i\text{mem}})$$

$$1/n_i F(\mu_{i\text{sol}} - \mu_{i\text{mem}}) = E^{\ominus} + \frac{RT}{n_i F} \ln a_i$$

So

$$\phi_{\text{mem}} - \phi_{\text{sol}} = E^{\ominus} + \frac{RT}{n_i F} \ln a_i$$

$$I_D = K(V_G - V_T - E_{\text{ref}} - \left(\frac{RT}{n_i F}\right) \ln a_i - V_D/2)V_D$$

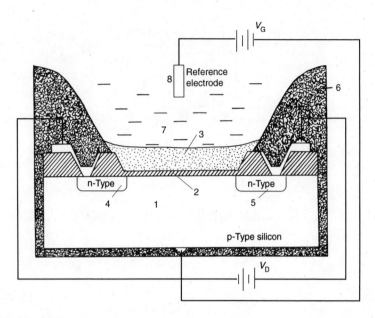

Figure 4.27 Diagram of the CHEMFET. 1, Silicon substrate; 2, insulator; 3, chemically sensitive membrane; 4, source; 5, drain; 6, insulating encapsulant; 7, analyte solution; 8, reference electrode

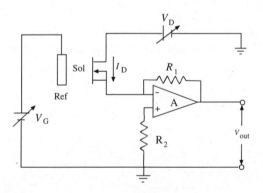

Figure 4.28 Schematic diagram of circuit for measuring I_D at constant gate voltage. A, Operational amplifier

where $K = C_O W \mu_n / L$

C_O = capacitance of insulator

W = width of channel

L = length of channel

μ_n = electron mobility

The current may be measured directly at constant V_G using a circuit such as that shown in Figure 4.28. Alternatively, one may keep I_D constant by changing V_G and measuring V_G using a circuit such as in Figure 4.29. This system is used in a number of sensor modes. The general CHEMFET mode has already been mentioned. The ISFET mode is the ion selective mode which uses the FET as an ion selective electrode. Following on from this, the ENFET is a biosensor in which the gate contains an enzyme system.

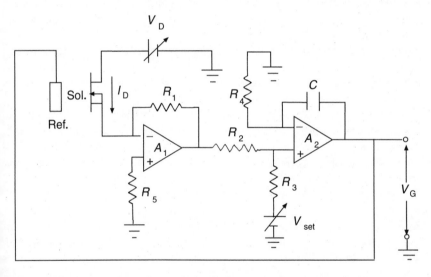

Figure 4.29 Schematic diagram of circuit for measuring changes in V_G at constant drain current. A_1, A_2 = Operational amplifiers

4.4.1 Applications of FET Sensors

4.4.1.1 CHEMFET

The simplest means of application is to use what is virtually a bare IGFET, with the gate consisting of a layer of palladium evaporated on to the silicon chip, and covered by a 100 nm oxide layer. This is highly specific for hydrogen gas down to 0.01 ppm. The response is $V = kp$ with $k = 27$ mV per ppm. There is some sensitivity of this electrode to CO, NH_3, H_2S, CH_4 and C_4H_{10}.

With the addition of a layer of iridium, the sensitivity to amonia is increased and that to hydrogen decreased.

4.4.1.2   ISFET

Ionophores are the most useful type of ion-selective polymer to use on FETs, as with ISEs. However, with FETs there are also some special responses for H^+. The first pH ISFET used the bare insulator gate as the ion-sensitive layer, but SiO_2 was not very effective, owing to the easy hydroxylation of the SiO_2. However silicon nitride (Si_3N_4) gate devices are not hydrolysed and are highly selective to H^+ ions, with a response of 50–60 mV per decade (pH). TiO_2 and Ge show a similar response. These semiconductor materials can be handled by the same techniques as used for FET chip fabrication. For other ions such techniques are less successful. However, for Na^+ ions borosilicate glass can be deposited in the gate region by integrated circuit processes.

Polymer membranes have been used successfully for K^+ incorporating a valinomycin crown ether and for Ca^{2+} with *p*-(1,1,3,3-tetramethyl-butyl)phenylphosphoric acid. Responses are <40 mV per decade, unless the membrane is thicker than 100 μm.

A quadruple function ISFET for H^+, Na^+, K^+ and Ca^{2+} has been described for clinical applications, using an Si_3N_4 bare gate for H^+, glass for Na^+, phosphoric acid derivative ionophore for Ca^{2+} and a valinomycin for K^+. It worked satisfactorily under laboratory conditions but with whole blood there were problems with the sodium analysis.

Heterogeneous membranes have had some success, the best being polyfluorinated phosphazine (PNF) mixed with silver salts, particularly silver chloride (75%) with PNF (25%). The response for Cl^- was 52 mV per decade. Changing the mixture and including Ag_2S or AgI can adjust the selectivity to favour a particular ion.

4.4.1.3   ENFETs: FET-based Biosensors

This is an excellent way of making a miniaturised biosensor. Usually a dual-gate arrangement is used, as in Figure 4.30.

The pH-based FET is the most often used, for example with penicillin and with glucose and urea. Indeed, a dual glucose–urea biosensor has been described, with three FET gates, one as a reference, one with glucose oxidase and one with urease.

The Pd-MOS FET device for hydrogen gas has been used with NAD^+/NADH and hydrogen-hydrogenase linked to pyruvate/NH_3, which with alanine dehydrogenase gives alanine. This suggests a whole range of possibilities for the many reactions involving NADH.

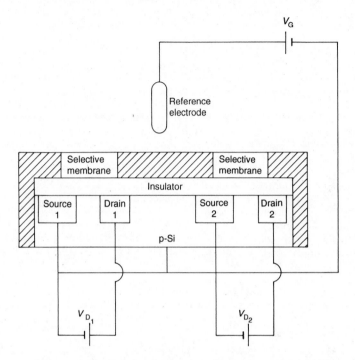

Figure 4.30 Schematic diagram of a dual-gate ENFET

$$\text{NADH} + \text{H}^+ \xleftarrow{\quad\text{hydrogenase}\quad} \text{H}_2 + \text{NAD}^+$$

$$\text{pyruvate} + \text{NH}_3 + \text{NADH} \xrightarrow{\quad\text{alanine dehydrogenase}\quad} \text{alanine} + \text{NAD}^+$$

The Ir-MOS device for ammonia has been used for urea, and a range of other analytes have been detected in a similar way, such as aspartate, asparginine, glutamate and creatinine.

4.5 REFERENCES

R. N. Adams (1969) *Electrochemistry at Solid Electrodes*, Marcel Dekker, New York.
W. J. Albery and D. H. Cranston (1987) 'Amperometric enzyme electrodes', in A. P. F. Turner, I. Karube and G. S. Wilson (Eds), *Biosensors: Fundamentals and Applications*, Oxford University Press, Oxford, Chap. 12, pp. 180–210.
A. J. Bard and L. R. Faulkner (1980) *Electrochemical Methods: Fundamentals and Applications*, Wiley, New York.
P. N. Bartlett (1987) 'The use of electrochemical methods in the study of modified

electrodes', in A. P. F. Turner, I. Karube and G. S. Wilson (Eds), *Biosensors: Fundamentals and Applications*, Oxford University Press, Oxford, Chap. 13, pp. 211–246.

J. J. Benschoten, J. Y. Lewis, W. R. Heineman, D. A. Roston and P. T. Kissinger (1983) *J. Chem. Educ.*, **60**, 772.

G. F. Blackburn (1987) 'Chemically sensitive field effect transistors', in A. P. F. Turner, I. Karube and G. S. Wilson (Eds), *Biosensors: Fundamentals and Applications*, Oxford University Press, Oxford, Chap. 26, pp. 481–530.

A. E. G. Cass, G. Davis, G. D. Francis, H. A. O. Hill, W. J. Aston, I. J. Higgins, E. V. Plotkin, L. D. Scott and A. P. F. Turner (1984) 'Ferrocene mediated enzyme electrode for the amperometric determination of glucose', *Anal. Chem.*, **56**, 657.

D. R. Crow (1994) *Principles and Applications of Electrochemistry*, Chapman and Hall, London, 4th edn.

E. A. H. Hall (1990) *Biosensors*, Open University Press, Milton Keynes.

J. Janata (1989) *Principles of Chemical Sensors*, Plenum Press, New York.

I. Karube (1987) 'Microbiosensors based on silicon technology fabrication', in A. P. F. Turner, I. Karube and G. S. Wilson (Eds), *Biosensors: Fundamentals and Applications*, Oxford University Press, Oxford, Chap. 25, pp. 471–480.

D. B. Kell (1987) 'The principles and potential of electrical admittance spectroscopy: an introduction', in A. P. F. Turner, I. Karube and G. S. Wilson (Eds), *Biosensors: Fundamentals and Applications*, Oxford University Press, Oxford, Chap. 24, pp. 427–468.

S. S. Kuan and G. Guilbault (1987) 'Ion selective electrodes and biosensors based on ISEs', in A. P. F. Turner, I. Karube and G. S. Wilson (Eds), *Biosensors: Fundamentals and Applications*, Oxford University Press, Oxford, Chap. 9, pp. 135–152.

T. C. Pierce, J. W. Gardner, S. Friel, P. N. Bartlett and N. Blair (1993) 'Electronic nose for monitoring the flavour of beer', *Analyst*, **118**, 371.

J. M. Slater, J. Payntor and E. J. Watt (1993) 'Multi-layer coating polymer gas sensor arrays for olfactory sensing', *Analyst*, **118**, 379.

Chapter 5

Transducers II—Optical Methods

5.1 INTRODUCTION

Most bioassays were originally of the photometric type. They involved changes in a species which involved a strong change in photometric properties. Probably the best known is the involvement of NAD⁺/NADH in biochemical reactions. For example:

$$\text{pyruvate} + \text{NADH} + \text{H}^+ \xrightarrow{\text{LDH}} \text{L-lactate} + \text{NAD}^+$$

NADH has a strong absorbance at λ_{max} 340 nm, but NAD⁺ has no absorbance at this wavelength. NADH also gives fluorescence at 400 nm. The absorption spectra of NADH and NAD⁺ are shown in Figure 5.1.

This may be used in the conventional way in which reagents and analyte are measured out and mixed and then placed in a cuvette in a spectrophotometer. The problem is how to make a sensor using an optical technique.

The basic optical response is based on the Beer–Lambert law (usually referred to as Beer's law):

$$\log(I/I_0) = A = \epsilon C l$$

where
- I_0 = intensity of incident light;
- I = intensity of transmitted light;
- A = absorbance (usually measured directly by instrument);
- ϵ = extinction coefficient;
- C = concentration of analyte;
- l = path length of light through solution.

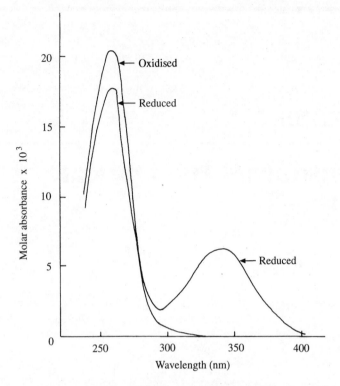

Figure 5.1 Absorption spectra of NAD in oxidised and reduced forms. Reproduced by permission of the Open University Press from Hall (1990)

The path length can limit the size of sensors, which does not occur in electrochemical devices.

The advantages of photometric devices are as follows:

(i)   No 'reference electrode' is needed—but a reference source is useful.

(ii)   There is no electrical interference.

(iii) An immobilised reagent does not have to be in contact with optical fibres. It can easily be replaced.

(iv)   There are no electrical safety hazards.

(v)   Some analytes, such as oxygen, can be sensed in equilibrium.

(vi)   They are highly stable with respect to calibration, especially if one can measure the ratio of intensities at two wavelengths.

(vii) They can respond simultaneously to more than one analyte using multiple immobilised reagents with different wavelengths for response, e.g. O_2 and CO_2.

(viii) Multi-wavelength measurements can be made to monitor changes in the state of the reagent.

(ix)   They have potential for a higher information content than electrical transducers.

Their disadvantages are as follows:
(i)   They will only work if appropriate reagent phases can be developed.
(ii)   They are subject to background ambient light interference. This may be excluded directly or using a modulation technique.
(iii) They have a limited dynamic range compared with electrical sensors— typically 10^2 compared with 10^6–10^{12} for ion-selective electrodes.
(iv)   They are extensive devices, dependent on the amount of reagent, hence they are difficult to miniaturise.
(v)   There are problems with the long-term stability of the reagents under incident light.
(vi)   Response times may be slow because of the time of mass transfer of analytes to the reagent phase.

5.2   OPTICAL TECHNIQUES

Let us survey the range of optical techniques that could potentially be used in such sensors. Figure 5.2 shows the electromagnetic spectrum, showing the relative wavelengths of important types of radiation such as visible, ultraviolet and infrared.

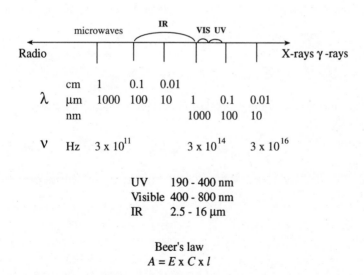

Figure 5.2   The electromagnetic spectrum

The main types of photometric behaviour which have been exploited in biosensors are as follows:

(i) Ultraviolet–visible absorption
(ii) Fluorescence (and phosphorescence) emission
(iii) Bioluminescence
(iv) Chemiluminescence
(v) Internal reflection spectroscopy (IRS)
(vi) Laser light scattering methods

5.3 OPTICAL FIBRES

Optical fibres have really only been developed over the past 10–15 years. They could be regarded as 'light conductors' or 'light wires'. Just as metal electrical wires will conduct electricity and bring power and electrical information to the point of need, often over very long distances, so optical fibres will do for light. Indeed, optical fibres are even now replacing electrical wires for telephone transmission. Optical fibres are *waveguides* for light. The original optical fibres were made of glass, but polymeric materials are now used, as they are much cheaper than glass and the metal wires used for electricity (Figure 5.3).

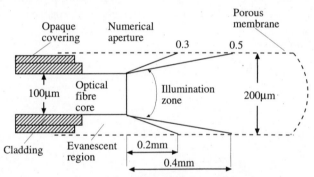

Figure 5.3 Basic structure of an optical fibre. Reproduced by permission of Oxford University Press from Schultz (1987)

The light waves are propagated along the fibre by *total internal reflection* (TIR). This is shown in Figure 5.4.

Total internal reflection depends on the angle of incidence and the refractive indices of the media.

$$\frac{\sin \theta}{\sin \phi} = \frac{n_2}{n_1} = n \text{ (Snell's Law)}$$

When $\phi = \pi/2$ and $\theta = \theta_c$, then $\theta_c = n_2/n_1$, where $\theta_c = $ critical angle for TIR. Thus, if $\sin \theta > n_2/n_1$ there is total internal reflection, but if $\sin \theta < n_2/n_1$ we have refraction and reflection.

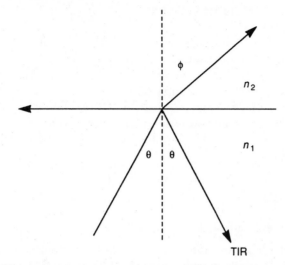

Figure 5.4   Illustration of Snell's Law and Total Internal Reflection

These principles need to be applied to a plane harmonic wave and to a cylindrical wave. One must consider the effect of change of phase angle and of bends in the fibre.

Waveguides in sensors are used in two ways, known as *extrinsic* or *intrinsic*. In the extrinsic mode, the waveguides simply act to transmit light from the light source to the light collector. In this mode, Beer's law can usually be used. In the intrinsic mode, the light is changed by the measurand: the phase, polarisation and intensity can be modulated within the waveguide by a measurand lying within the penetration depth for the evanescent field adjacent to the guide (Figure 5.5).

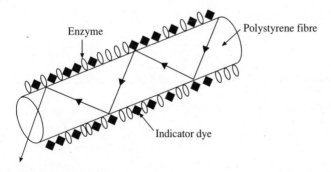

Figure 5.5   Model of an intrinsic enzyme sensor based on a polystyrene fibre. Reproduced by permission of the Open University Press from Hall (1990)

5.4   DEVICE CONSTRUCTION

No reference sensor is needed, unlike electrochemical transducers, for which there must always be a reference electrode. However, the system works better with a reference blank, the light source then being split between the sample and the reference. Detection may be effected at different analytical and reference wavelengths and one then obtains the ratio of the signals at the two wavelengths. This eliminates scatter and source fluctuations.

Waveguides for different forms of light may need to be made of different materials as follows;

$\lambda > 450\,$nm   plastic (such as polyacrylamide)
$\lambda > 350\,$nm   glass
$\lambda < 350\,$nm   fused silica
$\lambda > 1000\,$nm   germanium crystal guides.

In a photometric sensor, the reagent has to be immobilised so that it can interact with the analyte, probably to form a complex with distinctive optical properties which can be monitored by the sensor. This can happen in two main ways, as described in the following sub-sections.

5.4.1   Direct method

An analyte, A, can complex with the reagent, R, as follows:

$$A + R \rightleftharpoons AR$$

$$K = \frac{[AR]}{[A][R]}$$

If $\{R\}$ is the total reagent concentration, then

$$\{R\} = [R] + [AR]$$

and the free reagent concentration is

$$[R] = \frac{\{R\}}{1 + K[A]}$$

so the complexed reagent concentration is

$$[AR] = \frac{K[A]\{R\}}{1 + K[A]}$$

There is no simple relationship between the signal due to the free reagent and the analyte, or the combined reagent plus analyte.

If the measured optical parameter is proportional to [AR] then the response is proportional to [A] at low concentrations, such that $[A] \gg 1/K$; if $[A] \ll 1/K$ the response saturates to a limiting value.

If the response is proportional to [R], the signal decreases with increasing [A]. A linear relationship in this case can be obtained from

$$\{R\}/[R] = 1 + K[A]$$

A better approach, therefore, is to use two wavelengths, one for AR and one for R (reference). Now we can write

$$\frac{[AR]}{[R]} = K[A]$$

Hence there is now a linear relationship between the ratio of the absorbances at λ_{AR} and λ_R and the concentration of A. This relationship is independent of $\{R\}$.

5.4.2 Indirect Method

Sometimes a competitive binding method is used. This is particularly useful if neither the analyte nor the reagent show a spectral change on binding, in which case an *analyte analogue* may be used. This is a substance similar to the analyte which has inherent optical characteristics or which induces an optical change on bonding. The optical characteristic is often added to the analogue as, for example, a fluorescent label. In the determination of glucose, a suitable 'reagent', often called a bioreceptor, is concanvalin A (Con A), and a suitable analogue is dextran labelled with fluorescein isothiocyanate (FITC-dextran). The Con A is immobilised on a cellulose hollow fibre (see Figure 5.6).

The FITC-dextran is freely mobile inside the optical cell, but cannot diffuse out through the membrane. The analyte can diffuse through the membrane into the cell, because of the different relative sizes of the analyte and the analyte analogue (A*). In the absence of the sugar, the analyte analogue forms a complex with the receptor (R):

$$A^* + R \rightleftharpoons A^* \cdot R$$

The complex effectively removes some of the receptor from the walls of the fibre so that a signal equivalent to 20% of the maximum fluorescence is observed.

When the analyte (A) is introduced, there is a competing equilibrium with the analyte and the receptor:

$$A + R \rightleftharpoons A \cdot R$$

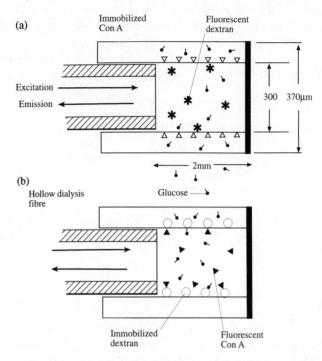

Figure 5.6 Two modes of biosensor for glucose using fluorescein isothiocyanate (FITC)-labelled dextran and Con A. Reproduced by permission of Oxford University Press from Schultz (1987)

This results in increasing amounts of receptor being displaced from the walls, resulting in further increases in the fluorescence signal. Eventually, all the receptor is free and no further increase in the fluorescence is seen: the signal is saturated. A typical response curve is shown in Figure 5.7.

One can analyse the behaviour as follows:

$$A + R \rightleftharpoons A \cdot R$$

$$A^* + R \rightleftharpoons A^* \cdot R$$

For simple unimolecular binding, the equilibrium constants for the two binding processes are

$$K = [A \cdot R]/[A][R]$$

$$K^* = [A^* \cdot R]/[A^*][R]$$

If we let the total amount of the receptor be {R} and that of the analyte analogue be {A*}, these total concentrations are conserved, so we can write

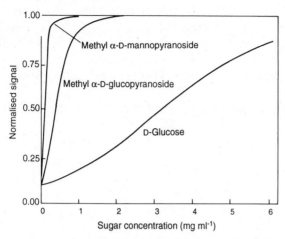

Figure 5.7 Response curves for different sugars for a Con A-based biosensor. Reproduced by permission of Oxford University Press from Schultz (1987)

$$\{R\} = [R] + [A·R] + [A*·R]$$

and

$$\{A*\} = [A*] + [A*·R]$$

to express the mass balances of the two reagents. These four equations can be solved to find the ratio of the bound receptor to the total receptor concentrations:

$$([A*]/\{A*\})^2 + \{A*\}/[A*]\left(\left(\{R\}/\{A*\} - 1\right) + \frac{K[A] + 1}{\{A*\}K*}\right) - \frac{K[A]+ 1}{[A*]K*} = 0 \quad (5.1)$$

$[A*]/\{A*\}$ represents the normalised response and has values between 0 and 1. The sensitivity is related to $\{A*\}$.

The normalised response is a function of two terms $\{R\}/\{A*\}$ and $(K[A] + 1)/\{A*\}K*$. The concentration of analyte $[A]$ occurs in the second term. Thus a complete characterisation of the model system can be obtained by a dimensionless plot of $[A*]/\{A*\}$ versus $(K[A] + 1)/\{A*\}K*$, for different values of $\{A*\}/\{R*\}$. This is shown in Figure 5.8.

These curves can be used to estimate the appropriate conditions for designing a biosensor, by the following procedure:

(i) Estimate the mid-range analyte concentration of interest, $[A']$.
(ii) Select a bioreceptor with a binding constant of the order of $10/[A']$.
(iii) Estimate the minimum concentration of A* that can be estimated with the optical system.$[A*']$, and assume $\{A*\}$ is about $50[A*']$.
(iv) Select (or synthesise) an analogue compound so that $K* = 1/[A*']$.
(v) Develop a technique for loading the sensor compartment with bioreceptor R such that the concentration of sites is about $100\{A*\}$.

Applying these principles to a glucose sensor, the range of blood glucose is 1–5 mg ml^{-1}, so [A′] is 2.5 mg ml^{-1} (0.025 M). K for Con A binding to glucose is 320 M^{-1}. The maximum amount of Con A that could be immobilised on the hollow surface of the dialysis fibre gave an effective concentration of 10^{-5} M. The binding constant K^* between FITC-dextran and Con A is about 7.5 × 10^4 M^{-1}. The total FITC-dextran concentration {A*} is about 1.5 × 10^{-6} M. The value of K[A′] was about 6, with glucose at a level of 2.5 mg ml^{-1}, and {R}/{A*} was 7. These data fit the above criteria.

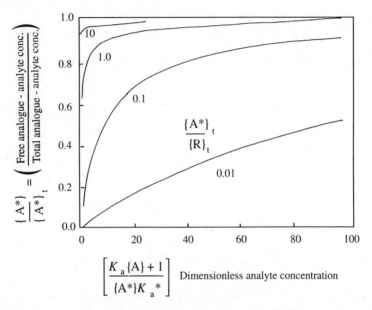

Figure 5.8 Parametric plot of equation 5.1. Changes in {A*}/{R} show the effect of relative levels of analyte-analogue concentration within the detection chamber. Reproduced by permission of Oxford University Press from Schultz (1987)

5.4.3   Solid-phase Absorption Label Sensors

The criteria are as follows:

(i)   Choice of immobilisation support.
(ii)   Immobilisation of indicators—retaining activity in desired rays.
(iii) Immobilisation of biorecognition molecules with retention of activity.
(iv)   Cell geometry.
(v)   Choice of source and detector components.

For a portable device, low power components are desirable, such as light-emitting diode (LED) sources and photodiode detectors. LEDs have only a limited range of wavelengths at present. The major ones are shown in Table 5.1.

The optical properties of an immobilised matrix are dependent on the cell geometry. For example, the indicator may be covalently bonded to polyacrylamide (PAA)-coated microspheres, resulting in much scattered light being detected.

Figure 5.9 shows a typical layout for a general system and Figures 5.10–5.13 show four modes of cell operation.

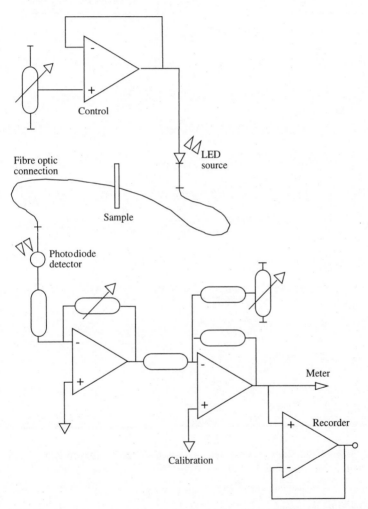

Figure 5.9 Basic control circuit for optical assay employing LED light source and photodiode detector. Reproduced by permission of the Open University Press from Hall (1990)

Table 5.1   LED light sources

Light source	Wavelength/nm	Light source	Wavelength/nm
Blue	455–465	*Detectors:*	
Yellow	580–590	Photodiode	560 (460–750)
Green	560–570		750
Red	635–695		800
Near-infrared	820		850
Infrared	930–950		900 (250–1150)
		Phototransistor	940

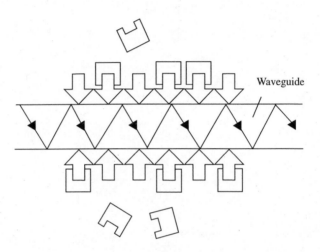

Figure 5.10   Optical sensor: evanescent field monitoring. Reproduced by permission of the Open University Press from Hall (1990)

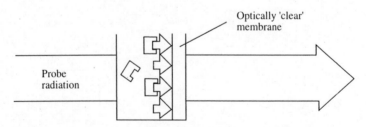

Figure 5.11   Solid-state optical sensor. Reproduced by permission of the Open University Press from Hall (1990)

5.4.4   Catalysis

The immobilised reagent can be a catalyst to convert the analyte into a substance with different optical properties. For example, immobilised

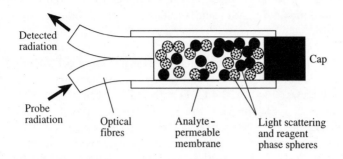

Figure 5.12 Optical sensor: light-scattering probe with common source and detection wavelength. Reproduced by permission of the Open University Press from Hall (1990)

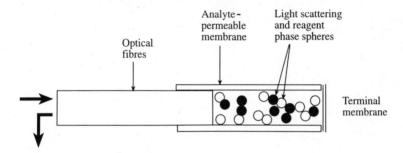

Figure 5.13 Optical sensor: light-scattering probe with different source and detection wavelengths. Reproduced by permission of the Open University Press from Hall (1990)

alkaline phosphatase catalyses the hydrolysis of *p*-nitrophenyl phosphate to *p*-nitrophenoxide, which has a strong adsorption spectrum.

5.5 SOME EXAMPLES OF POTENTIAL APPLICATIONS

5.5.1 Visible Absorption Spectroscopy

5.5.1.1 pH

Measurement of pH is fundamental to many measurements. Very many dyes act as pH indicators. The greatest problem with pH indicators is that the pH range of each one is relatively small—less than 2 pH units. However, this does not matter too much with biochemical systems as they generally operate within fairly limited pH ranges.

Methyl red is a suitable dye which has a distinctive visible spectrum with well separated maxima for the acid and basic forms of the dye. This is shown in Figure 5.14.

Methyl red can be incorporated into polyacrylamide-coated microspheres, which in turn were used in the system shown in Figure 5.11. Detection was originally made at 558 nm referenced to 600 nm, giving a pH range of 7.0–7.4 (± 0.01). A later version used an LED with 565 nm (bright green), referenced to an LED with λ 810 nm (infra-red). Detection was with a silicon p–i–n diode. This made an inexpensive portable pH sensor. One can see from Figure 5.14 that both wavelengths are well up towards the maximum for the acid form of methyl red and well away from the basic form. Both reference wavelengths are in the region of negligible absorbance of both species.

Carbon Dioxide

CO_2 can be determined using a pH probe with a gas-permeable membrane containing a hydrogencarbonate buffer.

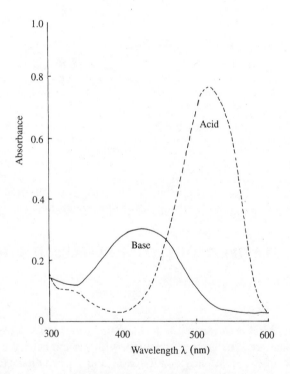

Figure 5.14 The ultraviolet–visible absorption spectrum of methyl red in its acid and base forms. Reproduced by permission of the Open University Press from Hall (1990)

Ammonia

A pH probe with a higher pH response is needed for ammonia. 4-Nitrophenol is a suitable indicator. A similar arrangement to the CO_2 sensor is used except that, of course, the buffer is ammonium chloride.

Others

A different approach has been tried using oxazine perchlorate dye, which responds directly to NH_3. It is coated on a capillary waveguide and operates in an intrinsic mode as shown in Figure 5.15. It is irradiated with a 560 nm LED and a photodiode detector is used.

The large range of $NAD^+/NADH$-related examples can also be included under this heading and also under the fluorescence mode.

5.5.1.2 EXAMPLES

A few examples have been developed into biosensors, e.g. for penicillin, urea and *p*-nitrophenyl phosphate. We shall look at penicillin as an example. Penicillinase catalyses the formation of penicilloic acid from penicillin with an increase in pH. A CNBr-activated optically clear cellulose membrane is covalently attached to the glutathione conjugate of bromocresol green. To this penicillinase is bonded via a carbodiimide reaction.

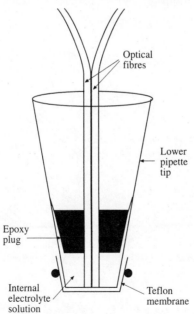

Figure 5.15 An optrode for ammonia. Reproduced by permission of the Open University Press from Hall (1990)

5.5.2 Fluorescent Reagents

This is perhaps the most developed area so far for photometric sensors and biosensors. One example has already been discussed in Section 5.4.2 for glucose with concanavalin A.

5.5.2.1 ION-SELECTIVE SENSORS

pH

Several fluorescent reagents have been developed for pH sensors. The best one is probably trisodium 8-hydroxy-1,3,6-trisulphonate. Absorption (excitation) bands occur at 405 nm (acid form) and 470 nm (basic form) with emission (fluorescence) at 520 nm, as shown in Figure 5.16. All these values are in the visible range, which enables less expensive optical ware to be used. It operates over the pH range 6.4–7.5 (± 0.01).

Halides

A fluorescence quenching sensor for Cl^-, Br^- and I^- has been designed using either acridinium- or quinidinium-based fluorescent reagents covalently bound to a glass support via carbodiimide. The best sensitivity was 0.15 mM for I^- using the acridinium sensor.

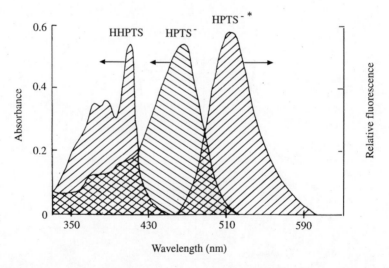

Figure 5.16 Absorption spectra of 8-hydroxy-1,3,6-trisulphonic acid (HPTS) in acid and base forms and fluorescent spectrum of base form. Reproduced by permission of the Open University Press from Hall (1990)

Na+

The optical label fluoro(8-anilino-1-naphthalene sulphonate) forms ion pairs with the immobilised ionophore–Na^+ complex in competition with a quenching cationic polyelectrolyte, poly[copper(II)–polyethyleneimine], which suppresses the fluorescence of excess label.

$$I-Na^+ + Poly-Fluor_{soln} \rightleftharpoons I-Na^+-Fluor + Poly_{soln}$$
$$\text{(quenched)} \qquad\qquad \text{(fluoresces)}$$

where I = ionophore.

K+

A photoactivated crown compound, selective for K^+, has been made by reacting 2-hydroxy-1,3-xylyl-18-crown-5 with diazotised 4-nitroaniline immobilised at the end of an optical fibre. Although the detection limit was 0.5 mM, the selectivity against Na^+ (= 1/6) was insufficient for clinical applications.

5.5.2.2 GAS SENSORS

Oxygen

Oxygen is a very efficient quencher of fluorescence, so this property can be used in a number of ways in sensors. One of the simplest is perylenebutyric acid on a polyacrylamide support, which is excited at 468 nm and fluoresces at 514 nm. This probe responds to oxygen over the range 0–150 Torr. An alternative sometimes used is pyrenebutyric acid, but it requires excitation at 342 nm, in the UV region, which means that inexpensive plastic fibre optics cannot be used. Another method, depending on fluorescence lifetime rather than intensity, has been developed for oxygen. It consists of tris(2.2′-bipyridyl)ruthenium(II) dichloride hydrate, which produces relatively long-lived fluorescence when excited at 460 nm to give fluorescence at 610 nm. The indicator was adsorbed on Kieselgel in a silicone membrane. Using a blue LED souce, the lifetime in the absence of oxygen was 205 ns. The sensor had an extended linear range and long-term stability.

Sulphur Dioxide

Benzo[*b*]fluoranthene responds down to 84 ppm of SO_2 in the absence of oxygen such as in exhaust gases, the arrangement being similar to that described for oxygen.

5.5.3 Chemiluminescence

Chemiluminescence occurs by the oxidation of certain substances, usually with oxygen or hydrogen peroxide, to produce visible light in the cold and in the absence of any exciting illumination, that is, in the dark. The best known of these is luminol, shown in Figure 5.17. There are other more complex substances which will react in a similar way.

The luminol is normally used as label. It can be used in any assay involving oxygen, hydrogen peroxide or peroxidase. It is particularly useful with immunoassays. Table 5.2 shows some examples of these. However, the sensitivity is limited because the quantum yield is only 1%.

A particularly interesting approach, combining luminescence and fluorescence, is shown in Figure 5.18. The antigen is labelled with the luminol

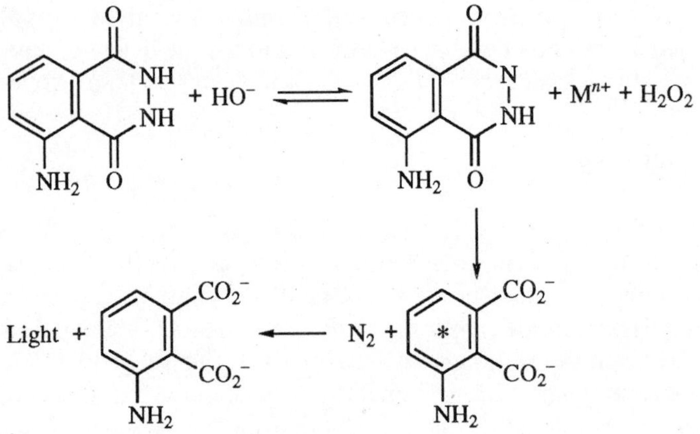

Figure 5.17 Reaction mode of luminol. Reproduced by permission of Oxford University Press from McCapra (1987)

Table 5.2 Examples of chemiluminescent immunoassays

Analyte	Label	Chemiluminescent reactants
Human IgG	Luminol	H_2O_2–haemin
Testosterone	Luminol derivative	H_2O_2–Cu(II)
Thyroxine	Luminol derivative	Microperoxidase
Biotin	Isoluminol	H_2O_2–lactoperoxides
Hepatitis B	Isoluminol derivative	Microperoxidase–peroxide
Rabbit IgG	Isoluminol	Microperoxidase–peroxide
Cortisol	Isoluminol	Microperoxidase–peroxide

Reproduced by permission of the Open University Press from Hall (1990)

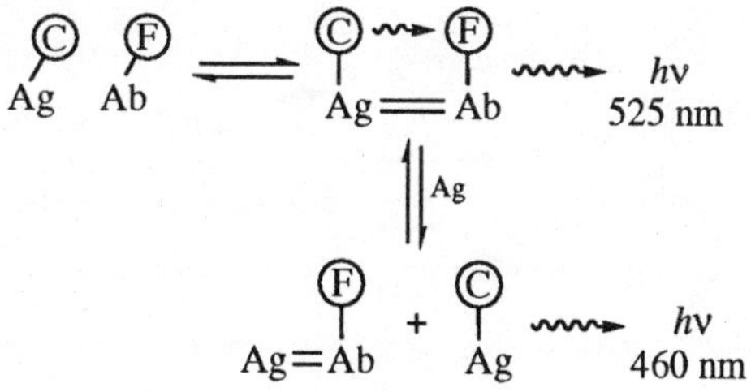

Figure 5.18 Competitive immunoassay employing a fluorescent-labelled antibody and chemiluminescent-labelled antigen. The fluorophore acts as the acceptor of the chemiluminescent energy in the doubly labelled complex. Reproduced by permission of Oxford University Press from McCapra (1987)

and the antibody is labelled with a fluorescent compound such that the emission from the luminol will excite the fluorescence.

In the bound Ag–Ab, the luminol emits light of 460 nm, which excites the fluorescer that in turn emits fluorescence at 525 nm, resulting in an increased quantum yield. At the same time, unlabelled antigen may combine with labelled Ab, in which case there is no fluorescencce so there is simply emission from the luminol at 460 nm. This permits analysis of both bound and unbound antigen at the two wavelengths.

Biosensors may be constructed involving luminol with hydrogen peroxide and peroxidase. A fibre optic sensor for H_2O_2 can be made with peroxidase immobilised on a polyacrylamide gel containing luminol at the end of the fibre optic. The luminescence is detected *in situ* without any diffusion of the luminescent species and, if $[S] \gg K_m$, it is independent of the thickness of the membrane. Of course, no external light source is needed. The sensor can be connected directly to the photodiode. It will detect 1–10 mM H_2O_2 with a response time of 2 min. One obvious application is to connect the sensor to a glucose–glucose oxidase reaction to determine glucose, for which a linear range of 0.15–1.5 mM can be obtained.

A recent example of a new type of chemiluminescence involves adamantyl dioxetine phosphate, shown in Figure 5.19, which can be hydrolysed under the influence of alkaline phosphatase to form adamantyldioxetaine anion, which is unstable and fluoreseces. The fluoresence lifetime is several minutes, unlike the conventional luminescence. The structures and fluorescence are shown in Figure 5.19.

This could be used in many types of assay which involve phosphate ester hydrolysis using alkaline phosphatase. The adamantyldioxetane phosphate could be allowed to compete with other organic phosphates for the phosphatase.

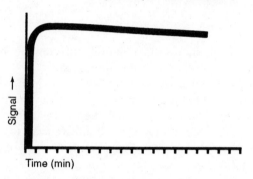

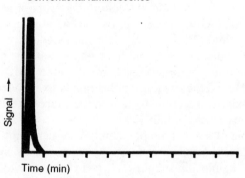

Figure 5.19 Mechanism of alkaline phosphatase-activated luminescence through hydrolysis of phosphate ester of adamantyldioxetine

5.5.4 Bioluminescence

Certain biological species, principally the firefly, can emit luminescence. This originates in a group of substances of varied structure called luciferins, one of which is illustrated in Figure 5.20.

The enzyme-catalysed oxidation of luciferin results in luminescence:

$$\text{luciferin} \xrightarrow{\text{luciferase, O}_2} \text{oxyluciferin} + h\nu \ (562\,\text{nm})$$

Some of the luciferase reactions couple with cofactors such as ATP, FMN and FADH, e.g.

$$\text{ATP} + \text{luciferin} + \text{O}_2 \xrightarrow{\text{luciferase}} \text{AMP} + \text{PP} + \text{oxyluciferin} + \text{CO}_2 + \text{H}_2\text{O} + h\nu$$

PP = pyrophosphate

This reaction is very sensitive down to femtomole concentrations. The determination of creatine kinase, which is related to the diagnosis of myocardial infarction and muscle disorders, is one important clinical assay that can be carried out by this method.

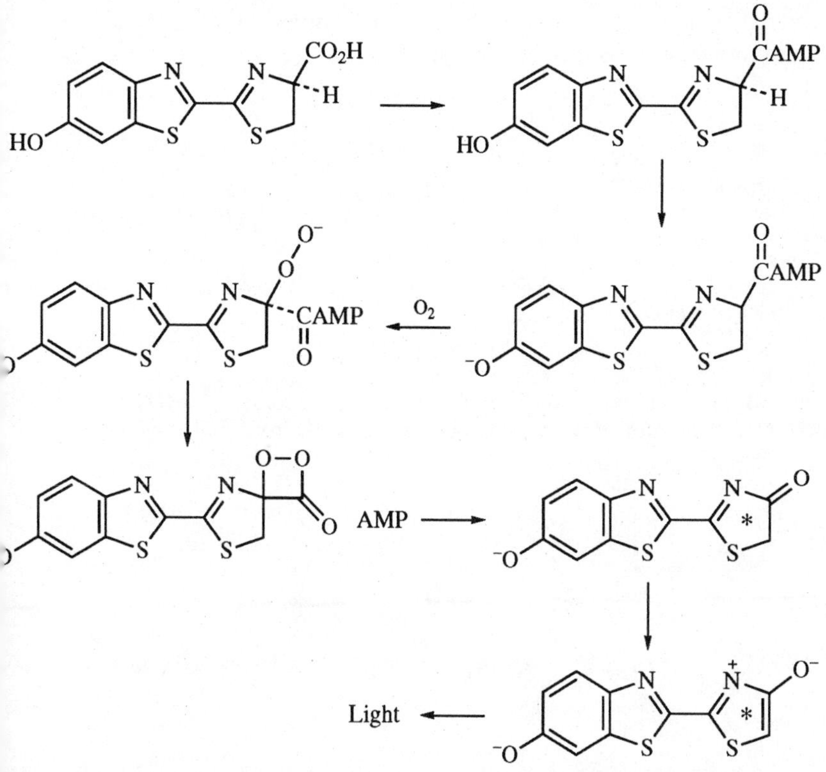

Figure 5.20 Examples of luciferins. Reproduced by permission of Oxford University Press from McCapra (1987)

$$AMP + \text{creatine phosphate} \xrightarrow{\text{creatine kinase}} ATP + \text{creatine}$$

$$ATP + \text{luciferin} + O_2 \xrightarrow{\text{luciferase}} AMP + PP + \text{oxyluciferin} + CO_2 + h\nu$$

Bacterial luciferases do not involve luciferins but form an excited complex with reduced flavins such as FMNH as follows:

$$FMNH_2 + O_2 + RCHO \longrightarrow FMN + RCOOH + H_2O + h\nu \ (478\text{–}505\,\text{nm})$$

Most analytical reactions involve NADH, as shown in Figure 5.21. This could be coupled, for example, to determine ethanol:

$$EtOH + NAD^+ \xrightarrow{\text{ADH}} CH_3CHO + DADH + H^+$$

as shown in Figure 5.22.

A very dramatic application is the determination of TNT using this system via an immunoassay, as shown in Figure 5.23.

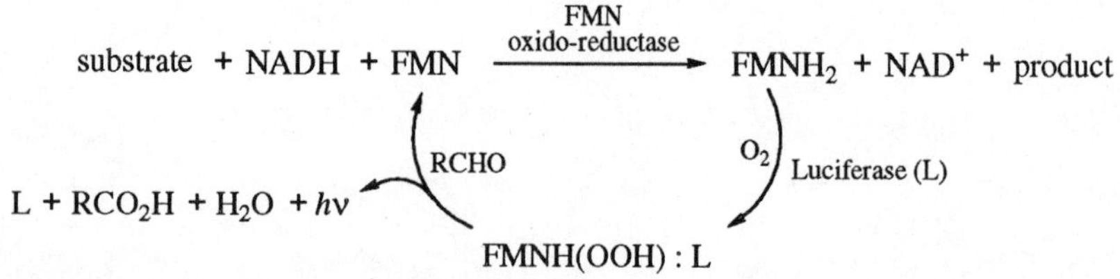

Figure 5.21 Basic mode of operation of bacterial luciferases via NAD and FMN. Reproduced by permission of the Open University Press from Hall (1990)

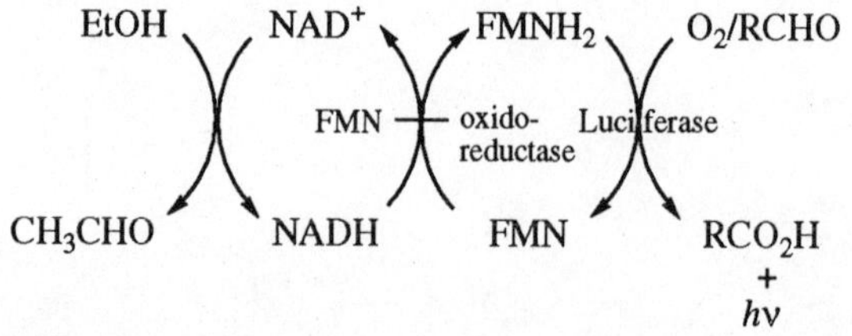

Figure 5.22 Bioluminescent assay of ethanol via NAD and ADH. EtOH/DAD/luciferins. Reproduced by permission of the Open University Press from Hall (1990)

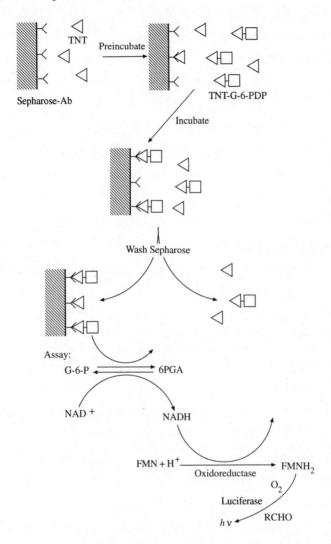

Figure 5.23 Amplified bioluminescent immunoassay of TNT. Reproduced by permission of the Open University Press from Hall (1990)

5.6 REFLECTANCE METHODS—INTERNAL REFLECTANCE SPECTROSCOPY (IRS)

These methods are concerned with studying material adsorbed on an optical surface. They are particularly suitable for use with immunoassays. There are three principle variations:

ATR—attenuated total reflectance;
TIRF—total internal reflection fluorescence;
SPR—surface plasmon resonance.

The basic arrangement is shown in Figure 5.24 (similar to Figure 5.4).

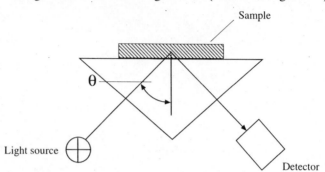

Figure 5.24 Attenuated total reflection. Reproduced by permission of Oxford University Press from Sutherland and Dähne (1987)

With a light wave striking the interface between two media of refractive indices n_1 and n_2, total internal reflection occurs when the angle of reflection θ is such that

$$\sin\theta_c = n_2/n_1 \text{ and } n_1 > n_2$$

If $\theta > \theta_c$, there is an evanescent wave refracted through the interface in the Z direction which penetrates the n_2 medium a distance d_p, which is of the order of a wavelength. It can be shown that the electrical field vector of this wave (E) is largest at the interface (E_0) and decays exponentially with distance (Z) as shown in the equation

$$E = E_0\exp(Z/d_p) \qquad (5.2)$$

This is illustrated in Figure 5.25 (a) and (b).

The depth of penetration, d_p, can be related to other factors by the equation

$$d_p = \frac{\lambda/n_1}{2\pi[\sin^2\theta - (n_2/n_1)^2]^{1/2}} \qquad (5.3)$$

One can see from equation 5.3 that qualitatively, d_p decreases with increasing θ and increases as n_2/n_1 tends to unity. One can thus select a value of d_p by an appropriate choice of values of θ, n_1 and λ. For example, with a quartz waveguide, $n_1 = 1.46$ and, if the sample is in water, $n_2 = 1.34$. This gives a

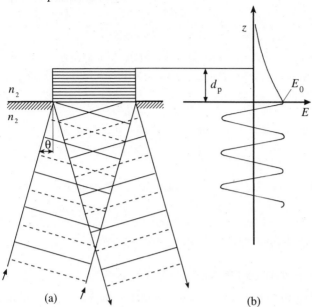

Figure 5.25 Generation of the evanescent wave at an interface between two optical media. Reproduced by permission of Oxford University Press from Sutherland and Dähne (1987)

value for θ_c of 66°. If θ is set to 70° and λ to 500 nm, the value of d_p is 270 nm, which will easily contain a monolayer of an immunological component with a diameter of about 25 nm.

However, d_p is just one of the four factors determining the attenuation of reflection. The others are the polarisation-dependent electric field intensity at the reflecting surface, the sampling area and the matching of the two refractive indices n_1 and n_2. An effective thickness d_e takes account of all these factors. It represents the thickness of film required to produce the same absorption in a transmission experiment.

In order to enhance the sensitivity, multiple reflections may be used as shown in Figure 5.26. If the number of reflections N is a function of length

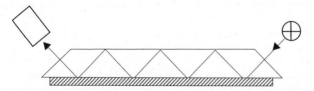

Figure 5.26 Total internal reflection showing multiple reflections. Reproduced by permission of Oxford University Press from Sutherland and Dähne (1987)

(L) and thickness (T) of the waveguide and the angle of incidence (θ), so that

$$N = (L/T)\cot\theta$$

then the longer and thinner the waveguide, the larger is N and the more frequently the evanescent wave interacts with the surface layer of analyte. If R is the reflectivity,

$$R = 1 - \alpha d_e$$

where α is the absorption coefficient. Then, for N reflections,

$$R^N = 1 - N\alpha d_e$$

5.6.1 Practical Arrangements for Internal Reflection Elements

Figures 5.27 and 5.28 show two practical arrangements for IREs. The first shows one with multiple reflections and the second one uses an optical fibre. Both types could be used for ATR or SPR methods.

5.6.2 Attenuated Total Reflection (ATR)

An absorbing material is placed in contact with the reflecting surface of an internal reflection element (IRE), causing attenuation of the internally reflected light. The intensity of the light is measured against the incident

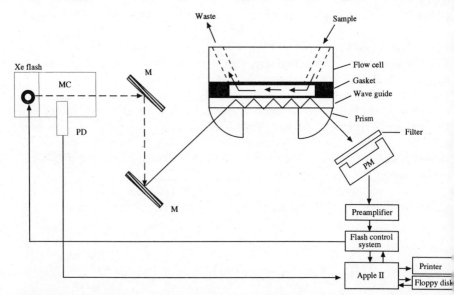

Figure 5.27 Diagram of instrumental layout used for measuring immunoassays with a multiple-internal reflection plate. PM, photomultiplier tube; PD, photodiode; MC, monochromator; M, mirrors. Reproduced by permission of Oxford University Press from Sutherland and Dähne (1987)

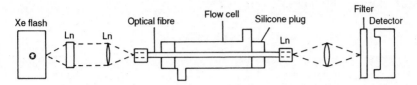

Figure 5.28 Diagram of fibre optic assembly with flow cell and light coupling optics for measuring immunoassays. In, lenses. Reproduced by permission of Oxford University Press from Sutherland and Dähne (1987)

wavelength. ATR has been used for monitoring immunoassays in the IR, visible and UV regions. For example, haemoglobin and rabbit anti-haemoglobin antisera have been monitored by ATR. The rabbit antibodies were covalently immobilised on the surface of a quartz microscope slide and then reacted with different concentrations of haemoglobin. The attenuation of the reflected light was measured at 410 nm. A similar application worked at 310 nm.

5.6.3 Total Internal Reflection Fluorescence (TIRF)

The emitted fluorescence can be detected either by a detector at right-angles to the interface (Figure 5.29a) or in line with the primary beam (Figure 5.29b). In fact, it can be shown theoretically and experimentally that the latter method results in an enhancement of up to 50-fold. This is particularly useful when using an optical fibre as the IRE. Also, by avoiding measurement of the fluorescence through the bulk of the sample solution round the fibre, interference is minimised. TIRF has also been used to measure immunological reactions. Phenylarsonic acid was immobilised on the surface of a quartz microscope slide via a hapten–albumin conjugate. The FITC-labelled antibody which could bind to the immobilised hapten could be detected by exciting fluorescence at the surface with the evanescent wave.

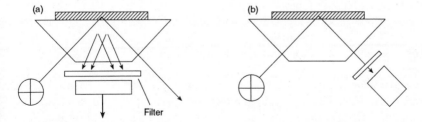

Figure 5.29 Detection of internal reflection by (a) right-angled fluorescence and (b) in-line fluorescence. Reproduced by permission of Oxford University Press from Sutherland and Dähne (1987)

5.6.4   Surface Plasmon Resonance (SPR)

5.6.4.1   BREWSTER ANGLE MEASUREMENTS

For plane polarised light, at the Brewster angle, which is defined as $\theta = \tan^{-1}(n_1/n_2)$, the reflectance $(R_p) = 0$. However, for a transparent incident phase and an adsorbing substrate n_2 is given by $N_2 = n_2 + ik_2$ $[i = (-1)^{1/2};$ $k_2 \neq 0]$ and R_p shows a minimum at the *pseudo*-Brewster angle. The angle for which R_p is a minimum $(R_{p\,min})$ is very sensitive to overlayers on the surface, so we can measure the change in R $(=\Delta R)$:

$$\Delta d_{min} = (\Delta R/\Delta d)^{-1}\Delta R_{min}$$

where d = thickness of surface layer. If ΔR_{min} is about $\pm 0.025\%$ then d_{min} is about 0.05 nm. If this is used for an antibody–antigen reaction of concentrations about 0.8 μg cm^{-2}., it has been suggested that the detection limit would be about 0.02–0.05 μg cm^{-2}, corresponding to a thickness change of 0.2–0.4 nm.

5.6.4.2   SURFACE PLASMONS

These are formed in the boundary of a solid (metal or semiconductor). The electrons behave like a quasi-free electron gas. Quanta of oscillations of surface charges are produced by exterior electrical fields in the boundary. These charge oscillations couple with high-frequency electromagnetic fields extending into space.

They can be excited by electron beams or by light. The most useful are non-radiative plasmons excited by evanescent light waves. The plasmon is characterised by an exponential decrease in the electric field with increasing distance from the boundary.

The normal technique for SPR uses the basic ATR configuration. The prism of refractive index n_1 is coated with a very thin layer of metal, such as 60 nm of silver, with a refractive index n_2, on to which is deposited a layer of sample of refractive index n_3 so that $n_1 > n_3$ as in Figure 5.30.

The p-polarised incident field, which has an angle θ such that the photon momentum along the surface, matches the plasmon frequency, so that the light can couple to the electron plasma in the metal. This is *surface plasmon resonance*.

The intensity of the totally reflected light is measured. It shows a sharp drop with increase in θ, depending on the depth and width and the characteristic absorbance and thickness of the metal. The effect can be seen in Figure 5.31, which shows the distribution of energy density $|H(x)|^2$ across the metal layer for different angles of incidence.

For $\theta_0 = 50°$, the energy is outside the resonance region and it decays

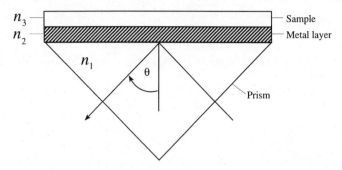

Figure 5.30 Attenuated total reflection method of exciting non-radiative plasmons using the Kretschmann arrangement. n_1, n_2 and n_3 are the refractive indices of the glass prism, metal layer and sample, respectively. Reproduced by permission of Oxford University Press from Sutherland and Dähne (1987)

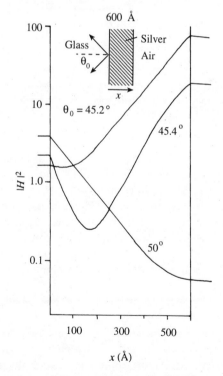

Figure 5.31 The calculated electromagnetic density $|H(x)|^2$ in a 600 Å silver film with the angle of incidence as the parameter. Reproduced by permission of Oxford University Press from Sutherland and Dähne (1987)

exponentially inside the plasma. For $\theta_0 = 45.4°$, it is near to the resonance and
the energy dips to a minimum and then rises to a higher value at the boundary.
When $\theta_0 = 45.2°$, resonance is reached, so the field energy rises to a maximum
at the boundary, with a value about 80 times the value without resonance.
Figure 5.32 shows a plot of intensity (R_p) versus angle of incidence (θ).

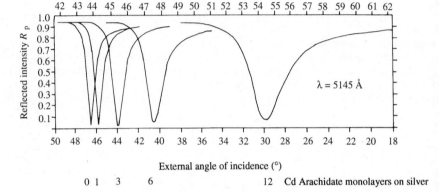

Figure 5.32 Attenuated total reflectivity curves for silver films covered with different
numbers of arachidate monolayers. Reproduced by permission of Oxford University
Press from Sutherland and Dähne (1987)

One can see that as the adsorbed layer becomes thicker the angle for
resonance shifts progressively to larger values and the width of each peak
increases. If the optical system can resolve better than 0.05°, coatings of a few
ångstroms thickness can be measured. The angular position is also very
sensitive to changes in refractive index just outside the metal film. Thus,
changing from air ($n = 1.0$) to water ($n = 1.33$) causes a shift in resonance
angle of 25°.

The experimental arrangement for SPR is similar to that for ATR, as
shown in Figure 5.30.

The IRE prisms, usually made of quartz, are coated with a thin layer of a
metal such as Au, Ag, Al or Cu. One can achieve the point of resonance in
several ways. One can keep the wavelength constant and vary the angle to
obtain maximum intensity of the reflected wave. Angular shifts as small as
0.0005° have been detected. An alternative is to keep the angle constant and
vary the wavelength to obtain the same effect. Light scattered by the surface
roughness of the film may be measured against the incident angle at fixed
wavelength, using a geometry as in Figure 5.29b.

The main applications of SPR so far have been in immunoassays. One of the immunological pair has to be immobilised to the IRE surface, which can be a tricky procedure. The most commonly used methods are adsorption and covalent bonding (see Chapter 3).

A model example is the reaction between human IgG and anti-IgG. The antigen is adsorbed on to the silvered surface of the prism to give a protein layer 50 Å thick. Various concentrations of anti-IgG were incubated with the adsorbed layer and the shift in resonance angle measured at fixed incident angle, as shown in Figure 5.33.

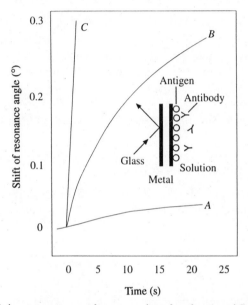

Figure 5.33 Shift in resonance angle versus time for three anti-IgG concentrations: (A)2, (B)20 and (C)200 μg ml⁻¹. The inset illustrates the antibody-binding event. Reproduced by permission of Oxford University Press from Sutherland and Dähne (1987)

This is a relatively new technique which has great potential. An SPR device called the BIAcore is marketed by Pharmacia Biosensors (Sollentuna, Sweden).

5.7 LIGHT-SCATTERING TECHNIQUES

The use of laser light has revolutionised the use of light scattering. There are a number of modes of scattering, such as Rayleigh scattering, Rayleigh–Gans–Debye scattering and Mie scattering. Many analytical techniques use light scattering. These can be divided into static and dynamic light scattering. The dynamic methods are more useful for applications involving biological materials. There are three new and powerful techniques in this category which we shall describe:

(i)   quasi-elastic light-scattering spectroscopy (QELS);
(ii) photon correlations spectroscopy (PCS);
(iii) laser doppler velocimetry (LDV).

5.7.1   Quasi-elastic Light-scattering Spectroscopy (QELS)

An assembly of particles suspended in a medium move under Brownian motion and the various light scatterings from each particle cause a fluctuation in the intensity with time. This distribution contains information about the size and size distribution of the particles. The time-scale of the intensity fluctuations is the basis of dynamic light scattering or intensity fluctuation spectroscopy, known as quasi-electric light-scattering spectroscopy (QELS). This has been used for the particle size determination of micelles and charged proteins.

5.7.2   Photon Correlation Spectroscopy

This technique is really an extension of QELS and is shown in Figure 5.34. The continuous laser light is passed through a cell containing the sample,

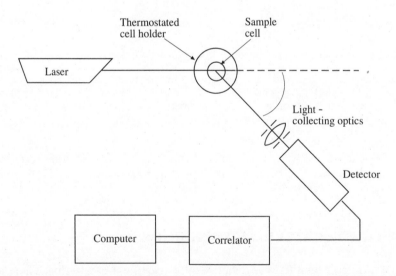

Figure 5.34 Layout of instrument for measuring photon correlation spectra. Reproduced by permission of Oxford University Press from Carr *et al.* (1987)

which will be in suspension under Brownian motion. The scattered laser light is collected by a lens and converted by a photodetector into an electrical signal. This signal is then analysed by a photon correlator. The correlator computes averages of the signal compared with itself at different delay times. This is called the autocorrelation function of the signal. It usually decays exponentially with the delay time. The delay time is related to the size of the scattered particles or macromolecules. From this function a particle size distribution is constructed. PCS operates with between 10^4 and 10^{10} particles/ml with particle sizes between nanometres and micrometres.

For immunoassays, PCS offers up to 100–1000-fold increases in sensitivity compared with conventional methods, and is comparable in sensitivity to radioimmunoassays, but without involving radioisotopes. There are many other applications to biomacromolecules, particularly with regard to the determination of the sizes and shapes of the particles, which can be of assistance in the determination of the tertiary structure and behaviour of biological macromolecules.

5.7.3 Laser Doppler Velocimetry

This technique is used to determine information about the velocity of particles flowing, for example, through a tube. The arrangement is shown in Figure 5.35.

Continuous plane-polarised laser light (a few mW power) is split into two equal beams, which are then focused to intersect in the fluid flow. The particles in the flow scatter the light from each laser beam with slightly different Doppler frequencies. Some of the scattered light is collected by a detector, causing beats at the photodetector. Analysis of the frequency data

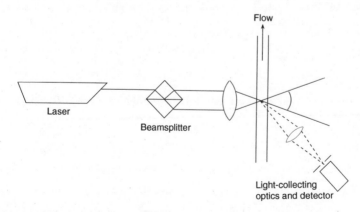

Figure 5.35 Laser Doppler velocimeter. Reproduced by permission of Oxford University Press from Carr *et al.* (1987)

received can give an estimate of the velocities of the particles. LDV has been applied to electrophoretic light-scattering studies of living cells, vesicles and counter-ion condensation on to DNA.

At present all the above techniques require large and bulky components and need frequency recalibration (except PCS), a high level of operator expertise and considerable sample preparation. However, with optical fibres it will be possible to simplify the procedures and permit the development of biosensors.

5.8 REFERENCES

L. J. Blum, S. M. Gautier and P. R. Coulet (1994) 'Fiber-optic biosensors based on luminometric detection', *Food Sci. Technol.*, **60**, 101.
R. J. G. Carr, R. G. W. Brown, J. G. Rarity and D. J. Clarke (1987) 'Laser light scattering and related techniques', in A. P. F. Turner, I. Karube and G. S. Wilson (Eds), *Biosensors: Fundamentals and Applications*, Oxford University Press, Oxford, Chap. 34, pp. 679–703.
D. Griffiths and G. Hall (1993) 'Biosensors—what real progress is being made?', *TIBTech.*, **11**, 122.
E. A. H. Hall (1990) *Biosensors*, Open University Press, Milton Keynes.
D. Leech (1994) 'Affinity biosensors', *Chem. Soc. Rev.*, 205.
F. McCapra (1987) 'Potential applications of bioluminescence and chemiluminescence in biosensors', in A. P. F. Turner, I. Karube and G. S. Wilson (Eds), *Biosensors: Fundamentals and Applications*, Oxford University Press, Oxford, Chap. 31, pp. 617–638.
M. Romito (1993) 'Biosensors: diagnostic workhorses of the future?', *S. Afr. J. Sci.*, **89**, 93, and references cited therein.
J. S. Schultz (1987) 'Design of fibre optic biosensors based on bioreceptors', in A. P. F. Turner, I. Karube and G. S. Wilson (Eds), *Biosensors: Fundamentals and Applications*, Oxford University Press, Oxford, Chap. 32, pp. 638–655.
W. R. Seitz (1987) 'Optical sensors based on immobilised reagents', in A. P. F. Turner, I. Karube and G. S. Wilson (Eds), *Biosensors: Fundamentals and Applications*, Oxford University Press, Oxford, Chap. 30, pp. 599–617.
R. M. Sutherland and C. Dähne (1987) 'IRS devices for optical immunoassays', in A. P. F. Turner, I. Karube and G. S. Wilson (Eds), *Biosensors: Fundamentals and Applications*, Oxford University Press, Oxford, Chap. 33, pp. 655–679.
P. Vadgama and P. W. Crump (1992) 'Biosensors: recent trends', *Analyst*, **117**, 1657.
R. H. Wilson, J. K. Holland and J. Potter (1994) 'Lining up for FTIR analysis', *Chem. Br.*, **30**, 993.

Chapter 6

Transducers III—Other Transducers

6.1 PIEZOELECTRIC CRYSTALS

Crystals have always had a fascination for people. Are not diamonds a girl's best friend? Their perfect shapes, their colours and the way they can be grown from a clear solution have long had a fascination. They have been (and are) involved in magic and sorcery and 'new age' religion. But they also have considerable scientific importance. The crystalline state is the one with the highest order, and therefore minimum entropy. Electrical properties of crystals have been of interest since the discovery of electricity. Crystal sets were the earliest form of radio transmitters and receivers. After many years in which thermionic valves dominated the radio industry, it was realised that crystals gave much more accurate frequency control than the capacitor–inductance coil combination. Now, of course, there has been a switch to transistors, but these sold-state devices are all crystals, as are all the 'chips' in computers.

The Curie brothers discovered in 1880 that anisotropic crystals, i.e. those having no centre of symmetry, such as quartz, tourmaline and Rochelle salt, give out an electrical signal when mechanically stressed. Conversely, if an electrical signal is applied to such crystals, they will deform mechanically. Thus, with an oscillating electrical potential the crystal will oscillate mechanically.

Each crystal has a natural reasonant frequency of oscillation, which can be modulated by its environment. The usual frequency is in the 10 MHz range, i.e. radio frequency. The actual frequency is dependent on the mass of the crystal together with any other material coated on it. The change in resonant frequency (Δf) resulting from the adsorption of an analyte on the surface can be measured with a high sensitivity (500–2500 Hz/g), resulting in sensors with picogram detection limits.

The relationship between the surface mass change, $\Delta m(g)$, and resonant frequency, f, is fairly straightforward and is given by the Sauerbrey equation:

$$\Delta f = -2.3 \times 10^6 f^2 \Delta m / A$$

where Δm is the mass (g) of adsorbed material on an area A (cm^2) of the sensing area. For a 15 kHz crystal a resolution of 2500 Hz/μg is likely, so that a detection limit of 10^{-12} g (1 pg) is probable.

The materials which show the piezoelectric effect and can be used in these devices now include (in addition to crystals) ceramic materials such as barium titanate and various lead zirconium titanates. Some organic polymers such as poly(vinylidene fluoride) (PVDF) ($-CF_2 -CH_2 -CF_2-$) also form crystals with piezoelectric properties.

The practicalities of this technique are shown in Figures 6.1 and 6.2.

It is often useful to use a differential mode with two balanced crystals and oscillators, as shown in Figure 6.3. This is called a quartz crystal microbalance (QCM).

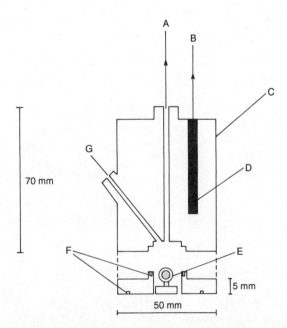

Figure 6.1 Cross-sectional view of piezoelectric crystal detector cell. A, Exit to waste; B, connection to temperature control box; C, stainless-steel detector cell; D, cartridge heaters; E, coated piezoelectric crystal; F, O-rings; G, butane and DMS inlet. Reproduced by permission of the Royal Society of Chemistry from Hawkesworth and Alder (1993)

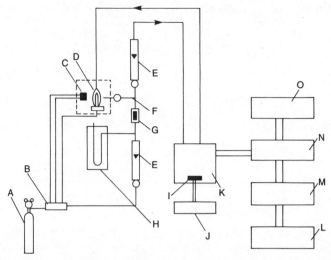

Figure 6.2 Schematic diagram of a test rig for a piezoelectric device. A, unstenched butane; B, pilot and cut-off switch; C, thermocouple; D, pilot flame; E, flow meter; F, isolation by-pass; G, injection port; H, manometer; I, cartridge heater; J, temperature control device; K, detector cell; L, recorder; M, frequency counter; N, oscillator circuit; O, power supply. Reproduced by permission of the Royal Society of Chemistry from Hawkesworth and Alder (1987)

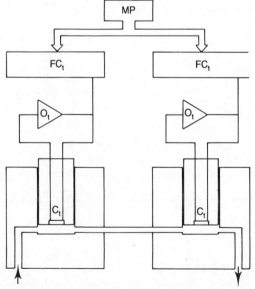

Figure 6.3 Typical system for electrogravimetric sensor analyses. Cr, Ct, References and test crystal sensors; Or, Ot, oscillating circuits; FCr, FCt, frequency counters; MP, microprotectors. Reproduced by permission of Oxford University Press from D. J. Clarke *et al.* (1987)

6.1.1   Examples of Applications

There has been much work with the analysis of gases, including SO_2, CO, HCl, NH_3 and CO_2. Specificity is obtained with selective coatings.

6.1.1.1   FORMALDEHYDE

A sensor has been developed for formaldehyde with formaldehyde dehydrogenase–NAD^+ as the selective layer. The enzyme is present in the dry state.

$$CH_2O + H_2O + NAD^+ \xrightarrow{\text{FDH}} NADH + HCO_2H + H^+$$

The enzyme and the glutathione cofactor were immobilised on a 9 MHz quartz crystal with glutaraldehyde cross-linking.

6.1.1.2   PESTICIDES (ORGANOPHOSPHORUS)

Cholinesterase enzymes can be used to detect organophosphorus pesticides, such as malathion. Acetylcholinesterase cross-linked with glutaraldehyde on to a quartz crystal responded to malathion in the gas phase. A 5 ppm concentration of water was needed to produce a satisfactory response, so the instrument had to be operated at constant humidity.

6.1.1.3   IMMUNOASSAYS

An antibody, anti-parathion immobilised on a quartz piezoelectric crystal, gives an alternative sensor for organophosphorus pesticides. Again, a constant humidity is needed.

6.1.1.4   SENSITIVE COATINGS

The crystal may be coated with an analyte-selective coating such as polypyrrole (as discussed in Chapters 4 and 5) coated on a thin gold film on the crystal surface. This has been used for the detection and measurement of a range of gases, including ammonia, methanol and hydrogen sulphide. In some other work a wide range of coatings was examined for the determination of dimethyl sulphide in liquid petroleum gas, and in another case for sulphur dioxide.

6.2   THE QUARTZ CRYSTAL MICROBALANCE

This is a very similar device which has received considerable attention recently. There have been problems due to variability, lack of sensitivity and inadequate signal-to-noise ratios. A proposed variation uses an amplification scheme referred to as amplified mass absorbant assay (AMISA). It has been used to measure adenosine-5'-phosphosulphate reductase (APS). The alkaline phosphatase-labelled antibody was bound to the microbalance surface, which was then exposed to 5-bromo-4-chloro-3-indolyl phosphate. This caused precipitation of an insoluble dephosphorylated dimer on the microbalance surface, enabling $5\,ng\,cm^{-3}$ of APS to be detected.

Other applications have been to *Salmonella typhimurium* and to a DNA strand of herpes simplex virus.

6.3 ELECTROCHEMICAL QUARTZ CRYSTAL MICROBALANCE (EQCM)

A thin layer of metal, such as gold or platinum, is plated on the surface of a piezoelectric crystal. This is made the working electrode in an electrochemical cell. The EQCM will detect changes in the mass of material at the electrode, such as (i) adsorption/desorption of species at the monolayer level, (ii) electrodeposition/electrodissolution at the electrode due to Faradaic redox processes and (iii) transfer of species between the solution and a surface-immobilised film.

A typical EQCM is made from a 10 MHz AT-cut quartz crystal, giving a sensitivity of $4\,ng\,cm^{-2}\,Hz^{-1}$, as shown in Figure 6.4.

It is preferable to compensate for changes in the properties of the solution, such as viscosity, which is affected by temperature, or density. The use of polymer-modified electrodes in sensors has been discussed in Chapters 3 and 4. The same principles can be applied here, except that the changes monitored are changes in mass measured by the quartz crystal microbalance. Some polymers used in this way include the redox polymers, polythionine, polybithiophene and polyvinylferrocene. EQCM has only recently been developed for use in biosensors.

6.4 ACOUSTIC WAVE MODES

These techniques use a piezoelectric crystal, particularly lithium niobate ($LiNbO_3$), but the waves are not generated in the bulk of the solution, but on the surface. A transmitter and a receiver are positioned on each end of the crystal, as shown in Figures 6.5 and 6.6.

The transmitter and receiver usually consist of sets of interdigitated electrodes. A radiofrequency applied to the transmitter produces a mechanical stress in the crystal, producing a Rayleigh-type surface acoustic wave which is received by the second set of electrodes and thus translated into an electrode voltage. The surface wave penetrates into the crystal to a

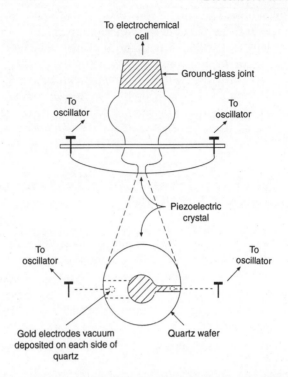

Figure 6.4 Electrochemical quartz crystal microbalance (EQCM). Adapted by permission of Blackie from P. A. Christensen and A. Hamnett (1994), *Techniques and Mechanisms in Electrochemistry,* Figs 2.108 and 2.109

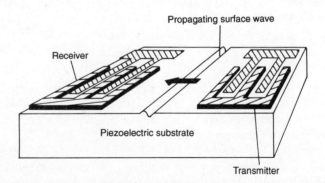

Figure 6.5 Model of surface acoustic wave (SAW) device. Reproduced by permission of the Royal Society of Chemistry from Zhang *et al.* (1993)

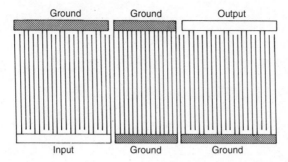

Figure 6.6 Schematic diagram of the metallised layout of the PMW transducer. Reproduced by permission of the Royal Society of Chemistry from Zhang *et al.* (1993)

depth of about one wavelength (rather like the evanescent optical wave; see Chapter 5). Thus a species immobilised on the surface will affect the transmission of the wave, unless the crystal is excessively thick, in which case the surface wave will be unaffected by a species on the underside of the crystal. There are a number of variations on this technique. Various acoustic wave modes include the following.

(*i*)   *Surface acoustic waves (SAW)*. These use Rayleigh surface waves as described above.

(*ii*) *The plate mode wave*. This involves waves being reflected through the bulk of the piezoelectric crystals to an interdigitated transducer (IDT) on one surface.

(*iii*) *The evanescent wave transducer*. This uses a substrate thickness less than the usual five acoustic wavelengths down to about 2–3 acoustic wavelengths in order to utilise the downward component (evanescent) of the wave. A ground plane is placed between the two IDT structures to prevent interference signals.

(*iv*) *The Lamb mode device*. This uses a very small substrate thickness compared with the wavelength. The excitation frequency is much lower. Often there is an analytically sensitive polymer layer on the lower surface, as shown in Figure 6.7.

This mode offers higher sensitivity and greater flexibility of sensor design. It has been used with a copper phthalocyanine coating for vapour analysis.

Although a considerable amount of work has been devoted to the study of properties of these devices, often using glycerol solutions, very few practical applications involving potential biosensors have been realised, although a number are under investigation, particularly for antibodies and proteins.

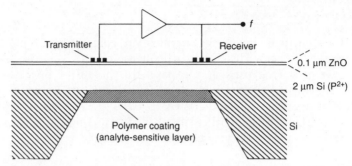

Figure 6.7 Transverse section through a Lamb mode device. Reproduced by permission of the Open University Press from Hall (1990)

The influenza virus has been detected on a SAW device in which the quartz surface was etched and coated with (3-glycidyloxypropyl) triethylsilane, followed by immobilisation of an antibody against the influenza virus. Exposure of the sensor to the virus resulted in frequency shifts.

In another system, a DNA probe was made using iron oxide particles as a mass amplifier. Entities bearing one of the reacting pairs were attracted to the sensor surface in a magnetic field.

(v)   *Thickness shear mode device (TSM).* This mode (Thompson *et al.*, 1991) has been used mainly with the device exposed to liquids. Extra variables are involved such as viscosity, density and conductivity of the liquid. The precise mode of operation of these devices is still open to question. It appears that only two types of analytical signals are generated by shear wave devices. First, thin films characterised by a shear modulus of elasticity give rise to standard Sauerbrey mass measurement. Second, capture at a liquid–solid shearing surface could lead to a differential signal associated with the introduction of new material at the interface. Applications have been made of this device towards biosensor development in several ways.

Candida albicans microbes were detected by means of the anti-*Candida* antibody covalently bonded on to plated platinum electrodes.

Human IgG was measured on an AT-cut 9 MHz crystal modified with protein A immobilised on an oxidised palladium layer on the crystal surface with (3-aminopropyl)triethoxysilane. Shifts in frequency were ascribed to the affinity of protein A for human IgG.

Other work has been done with immunoglobulins and protein A.

Surface plasmon resonance and IRS are similar techniques in that they observe the behaviour of surfaces, which have already been dealt with in Chapter 5.

6.5 THERMISTOR SENSORS

Biochemical reactions involving enzymes, in common with most other reactions, generate heat. This heat can be measured and, with the selectivity of the enzyme, can be used as the basis for thermistor biosensors. The technique requires a very sensitive electrical resistance thermometer, which will detect temperature changes down to 0.001°C. The usual amount of enthalpy change for biochemical reactions is 25–100 kJ mol^{-1}. Table 6.1 gives some molar enthalpies of enzyme-catalysed reactions.

Table 6.1 Molar enthalpies of enzyme-catalysed reactions

Enzyme	Substrate	$-\Delta H$/kJ mol^{-1}
Catalase	Hydrogen peroxide	100
Cholesterol oxidase	Cholesterol	53
Glucose oxidase	Glucose	80
Hexokinase	Glucose	28
Lactate dehydrogenase	Na pyruvate	62
NADH dehydrogenase	NADH	225
Penicillinase	Penicillin G	67
Trypsin	Benzoyl-L-arginineamide	29
Urease	Urea	61
Uricase	Uric acid	49

Reproduced by permission of Oxford University Press from Danielsson and Mosbach (1987)

The apparatus is shown in Figure 6.8. It consists of a temperature-controlled aluminium cylinder set to 25, 30 or 37°C in a polyurethane foam insulating cover. Another aluminium cylinder inside is linked up with the heat exchanger system to provide very stable temperature control. Into this is inserted the enzyme column, to which is connected the thermistor attached to a short gold capillary tube. The thermistor is typically a 16 kohm resistance (at 25°C) with a coefficient of −3.9%/°C. There is usually also a reference thermistor. The resistance is measured by a D.C. Wheatstone bridge which is amplified to give an output of up to 100 mV per 0.001°C. A typical sample containing 0.5–1 mM will give a response of about 0.01°C. The enzyme column is typically 30 × 7 mm i.d. nylon tubing or controlled-pore glass. The column needs about 100 units of enzyme, but can be used many times and lasts several months. Enzyme columns can be easily changed for a different assay. The sample is injected into a flowing buffer solution which flows through the enzyme column and out to waste. The buffer flow-rate should be between 0.5 and 3 ml min^{-1}, with a sample size of 0.1–1 ml. A temperature spike is observed which is directly proportional to the substrate concentration. The response is linear

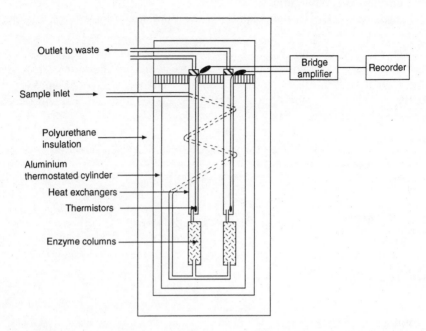

Figure 6.8   Thermistor device. Adapted by permission of Oxford University Press from Danielsson and Mosbach (1987)

over the range 10^{-5}–10^{-1} M and between 15 and 60 samples can be analysed per hour.

By optimising the enzyme support matrix and using improved heat exchangers in flow systems, sensitivities down to 0.01 mM have been achieved. The use of enzyme sequences to recycle the substrate or the coenzyme can have an amplification effect extending the assay into the nanomole range. Some say that these thermistor-type sensors are only 'honorary' biosensors! However, the thermal enzyme probe (TEP) with its close aposition of the enzyme to the thermistor surface conforms more closely to the 'true' biosensor concept. A thermoelectric glucose sensor used a thin-film thermopile on Mylar with a glucose oxidase–catalase combination (Vadgama and Crump, 1992).

6.5.1   Applications

This technique has been used in a wide range of areas, including clinical chemistry, determination of enzyme activity, monitoring gel filtration chromatography, process control and fermentation, environmental analysis and enzyme-linked immunoassays.

Some examples of enzyme-catalysed reactions which have been used are shown below:

$$\text{ascorbic acid} \xrightarrow{\text{scorbate oxidase}} 50\text{--}100\,\mu M$$

$$\text{cholesterol} \xrightarrow{\text{cholesterol oxidase}} 30\text{--}150\,\mu M$$

$$\text{glucose} \xrightarrow{\text{GOD--catalase}} 0.2\text{--}700\,\mu M$$

$$\text{phenol} \xrightarrow{\text{tyrosinase}} 1000\text{--}10\,000\,\mu M$$

$$\text{sucrose} \xrightarrow{\text{invertase}} 50\text{--}100\,000\,\mu M$$

6.6 REFERENCES

F. Benmakroha, T. Boudjerda, R. Boufenar, H. Allag, F. Djerboua and J. J. McCallum (1993) 'Monitoring of sulphur dioxide using a piezoelectric crystal based controller', *Analyst*, **118**, 401.

S. Bruckenstein and S. Shay (1985) 'Applications of the quartz crystal microbalance in electrochemistry', *J. Electroanal. Chem.*, **188**, 131.

P. A. Christensen and A. Hamnett (1994) 'The electrochemical quartz crystal microbalance', in *Techniques and Mechanisms in Electrochemistry*, Blackie, Glasgow, Chap. 2.2.6.

D. J. Clark, B. C. Blake-Coleman and M. R. Calder (1987) 'Principles and potential of piezo-electric tranducers and acoustical techniques', in A. P. F. Turner, I. Karube and G. S. Wilson (Eds), *Biosensors: Fundamentals and Applications*, Oxford University Press, Oxford, Chap. 28, pp. 551–572.

B. Danielsson and K. Mosbach (1987) 'Theory and application of calorimetric sensors', in A. P. F. Turner, I. Karube and G. S. Wilson (Eds), *Biosensors: Fundamentals and Applications*, Oxford University Press, Oxford, Chap. 29, pp. 575–597.

E. A. H. Hall (1990) *Biosensors*, Open University Press, Milton Keynes.

K. Hawkesworth and J. E. Alder (1993) 'Evaluation of coating materials used on piezoelectric crystals for the detection of dimethyl sulphide in liquified petroleum gas'. *Analyst*, **118**, 395.

D. Leech (1994) 'Affinity biosensors', *Chem. Soc. Rev.*, 205.

M. Thompson, A. L. Kipling, W. C. Duncan-Hewitt, L. V. Rajakovic and B. A. Cavic-Vlasak (1991) 'Thickness-shear-mode acoustic wave sensors in the liquid phase', *Analyst*, **116**, 881.

P. Vadgama and P. W. Crump (1992) 'Biosensors: recent trends. A review', *Analyst*, **117**, 1657.

P. W. Walton, P. M. Gibney, M. P. Roe, M. J. Lang and W. J. Andrews (1993)

'Potential biosensor system employing acoustic impulses in thin polymer films', *Analyst*, **118,** 425.

D. Zhang, G. M. Green, T. Flaherty and A. Shallow (1993) 'Development of interdigitated acoustic wave transducers for biosensor applications', *Analyst*, **118,** 429.

Chapter 7

Performance Factors

As with any relatively new technique, one needs to establish fairly quickly criteria by which its performance can be measured. As the technique is developed, these criteria have to be refined continuously as expectations are raised. This is especially true for a device containing biological material. Showing that a method will work (sort of!) in a laboratory is a very long way from delivering a commercial product into the hands of (say) a medical laboratory technician.

A number of factors will be discussed in this chapter which apply to biosensors. These will be illustrated by reference to examples that have largely been encountered already in this text. Unfortunately, different workers express their factors in different ways and, even worse, some, including a few of those one considers to be the best workers, do not give much data at all about performance in their published papers. This is probably so that they can protect their discoveries with patents, in case they can develop commercially viable biosensors. However, some data must be published as soon as possible for the benefit of other workers.

Anyone developing a new sensor of any sort needs to have some idea of what performance requirements are needed for the particular application in mind. Some of this sort of information is referred to in Chapter 10, but for illustration we give in Table 7.1 some requirements for environmental field applications.

7.1 SELECTIVITY

This factor is of the essence of biosensors; it is their *raison d'être*. It is established by the nature of the biological material. There are effectively three distinct types: enzymes, antibodies and nucleic acids. By far the most attention has been given to enzymes.

Table 7.1   General requirements for biosensors in environmental field applications

Requirement	Specification
Cost	£0.7–10 per analysis
Portable	Can be carried by one person, no external power required
Field use	Easily transported in van or truck; limited need for external power
Assay time	1–60 min
Personnel training	Can be operated by minimally trained personnel after 1–2 h training
Matrix	Minimal preparation for ground water, soil extracts, blood or saliva
Sensitivity	ppm–ppb
Dynamic range	At least two orders or magnitude
Specificity	Enzymes/receptors: specific to one or more groups of related compounds Antibodies: specific to one compound or one group of closely related compounds

One must ask how wide a range of chemical species will respond to a particular sensor. A sensor may be so specific that only one compound in one stereoisomeric structure causes any response. This sounds ideal, but it can raise its own problems. Alternatively, a broad range of similar chemical types may cause a response to some extent or another. This may be due to the nature of a very specific enzyme or it may be that the biological preparation actually contains a mixture of enzymes of similar specificity. Microorganisms are particularly likely to contain a mixture of enzymes, as are tissue preparations.

These interferences from other substances are much less likely to be a problem than with purely chemical sensors such as ion-selective electrodes.

7.2   RANGE AND LINEAR RANGE

With any analytical technique it is vital to know what concentration range of substrate is covered and, from the calibration point of view, over what section of this range the response is linear. At the lower limit is the detection limit. This has a precise definition according to the IUPAC convention. This is illustrated in Figure 7.1. The detection limit is defined by IUPAC as the

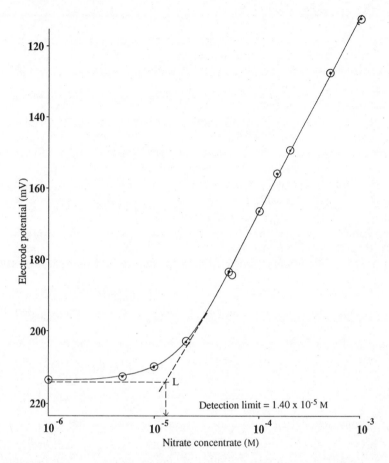

Figure 7.1 Calibration graph for nitrate showing the detection limit

concentration of the analyte at which the extrapolated linear portion of the calibration graph intersects the baseline—a horizontal line corresponding to zero change in response over several decades of concentration change. This is point L in Figure 7.1.

To be useful, the detection limit generally needs to be better than 10^{-5} M (0.01 mM). The importance of the range can be illustrated from the example of the determination of blood glucose. It needs to be at least 0.2–20 mM $(2 \times 10^{-4} – 2 \times 10^{-2}$ M) (preferably $1 \times 10^{-4} – 5 \times 10^{-2}$ M) to cover the likely blood glucose levels found in normal and diabetic patients.

7.3 REPRODUCIBILITY

This is always an important factor with any analytical technique, but especially so with biosensors, where it is impossible to reproduce the quality of biological preparations as well as with ordinary chemical substances. However, the quality control exercised by manufacturers and suppliers of purified enzymes (such as Alpha and Sigma) gives a reasonable idea of the number of active units (U) in an enzyme preparation. Unfortunately, most enzymes are unstable to some extent and after keeping for even short periods lose some activity. Therefore, it is generally necessary to check whether there is sufficient activity remaining by using a standard enzyme assay technique. This can often be done by simple spectroscopic assay as with glucose oxidase. The FAD in glucose oxidase has a λ_{max} at 450 nm. The absorbance at this wavelength is directly proportional to the glucose oxidase activity. For some enzymes the required assay may be more complex.

No analytical result has any real significance unless one can specify an estimate of the probable error. This applies even more so with biosensors. Several readings may be made for each determination, or much better, several replicate determinations can be made. Then a standard deviation may be calculated. This can usually be done on a pocket scientific calculator. If a calibration graph is to be plotted, the data may be fitted by a linear least-squares fit to give the best (i.e. calculated), straight line with its standard deviations and correlation coefficient (i.e. goodness of fit). This can be done with a simple computer program which is now often found in modest scientific calculators. More ambitious but very important is to correlate one set of data with another. Again, standard statistical methods are available to determine the relative standard deviation. With biosensors the expected reproducibility between replicate determinations should be at least ±5–10%.

Another way of avoiding the necessity for plotting a full calibration graph every time is to use the standard addition method, or, better, the multiple standard addition method. The sample to be analysed is measured, then a sample of a known standard with a concentration about twice that expected for the unknown is added and a second measurement is made. This method assumes that the biosensor response is known to vary linearly with sample concentration.

Let the response, r, be

$$r = k \times \text{concentration } (C); r = kC \tag{7.1}$$

For the unknown (u):

$$r_u = kC_u \tag{7.2}$$

For the sample with added standard (s):

$$r_{(u + s)} = k(C_u + C_s) \tag{7.3}$$

Dividing equation (7.3) by equation (7.2):

$$r_{(u + s)}/r_u = k(C_u + C_s)/kC_u$$

which equals $1 + C_s/C_u$. Rearranging gives

$$C_u = r_u C_s/[r_{(u + s)} - r_u]$$

With the multiple standard addition method, several aliquots of the standard are added. A graph is then plotted of response for each addition against amount of each addition. The line is then extrapolated back to zero response. The negative intercept on the concentration axis is the unknown concentration. This is shown in Figure 7.2.

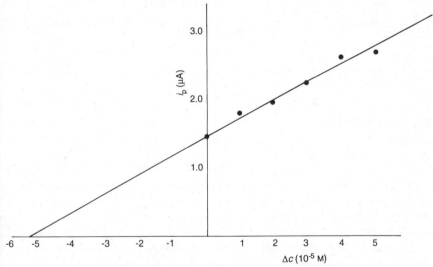

Figure 7.2 Multiple standard addition determination of catechol in beer

If the relationship between the analyte concentration and the sensor response is logarithmic (as with most potentiometric-based sensors), it is best to take the antilog and make a Gran plot. If

$$E = K + S_{log}(C_u + C_s)$$

then

$$10^{E/S} = 10^{K/S}(C_u + C_s)$$

This time one plots $10^{E/S}$ versus C_s; the negative intercept at $10^{E/S} = 0$ gives C_u, in a similar way to the multiple standard addition plot.

7.4 RESPONSE TIME

Many analytical devices require some settling down time, i.e. time to allow the system to come to equilibrium. This is true to some extent with many chemical methods; for example, our recent extensive studies with a nitrate electrode show that the most reproducible results are obtained after stirring the test solution in contact with the electrode for 30 s. With biosensors these measurement times may be longer than this. Obviously, if these times become too long it can materially affect the usefulness of the method for repetitive routine analyses. However, with sensors of a chemical or biochemical nature this response time is usually greatly offset by the simplicity of the measurement and the minimal sample preparation time. For biosensors, response times can vary from a few seconds up to several minutes. Up to 5 min is probably acceptable, but if the time exceeds 10 min, it may be too long.

Very much related to the response time is the recovery time—the time before a biosensor is ready to be used for the next sample. They may be the same thing or it may be that after one measurement the biosensor system has to 'rest' to resume a base equilibrium before it can be used with the next sample. In many papers these times are combined and the result is given as the number of samples that can be analysed per hour, which is what is really important in the end.

7.5 LIFETIME

As we know, all organic material deteriorates with time, especially when taken out of their natural environment. This means that one of the major drawbacks with biosensors is that the biological component usually has a fairly limited lifetime before it needs replacing. All developments of new biosensors include studies to show how the response of the biosensor to a standard sample changes with time over hours, days and months. Generally, pure enzymes have the lowest stability, whereas tissue preparations have the longest. One can consider three aspects of lifetime: the lifetime of the biosensor in use, the lifetime of the biosensor in storage and the lifetime of the biological material stored separately.

A number of techniques are currently being developed to improve this aspect of biosensors. Gibson *et al.* (1992) and Gibson and Woodward (1992) have succeeded in stabilising a range of enzymes by adding a mixture of a polyelectrolyte [diethylaminoethyl (DEAE)-dextran] and a sugar alcohol (lactitol) as soluble additives to enzyme solutions. This treatment enhanced the retention of enzyme activity in solution, during desiccation and during thermal stress. The principal enzymes studied were alcohol oxidase and horseradish peroxidase. Enhanced performance was also obtained with twelve other enzymes.

Further studies have been made with biosensors constructed for alcohol, using alcohol oxidase, one with membrane immobilisation and amperometric oxidation of hydrogen peroxide and the other with a mediated coupled reaction with horseradish peroxidase and NMP–TCNQ on a graphite electrode. In both cases addition of the stabilisers promoted a considerable increase in storage stability of the enzyme component, as indicated by an increase in the shelf life when stored in the dried form at 37°C. Similarly, an L-glutamate biosensor made from NMP–TCNQ-modified graphite electrodes and L-glutamate oxidase also showed an increase in shelf life when stored desiccated in the presence of stabilisers.

Other workers who attempted to use just the DEAE-dextran polyeletrolyte with glucose oxidase biosensors, with dimethylferrocene in carbon paste electrodes, were less successful. In fact, lyophilised glucose oxidase alone is extremely stable: it is stable for 2 years at 0°C and 8 years at −15°C. In solution it is most stable at pH 5. below pH 2 and above pH 8 the catalytic activity is rapidly lost.

7.6 EXAMPLES

7.6.1 Different Biomaterials

One of the best studied systems is a biosensor for glutamine based on glutaminase found in pork liver. This enzyme catalyses the deamination of glutamine to glutamate. The ammonia is detected potentiometrically with an ammonia-sensitive electrode. The enzyme is also found in mitochondria and in bacteria. A comparison was made of the various response features for a biosensor made with (i) the isolated enzyme, (ii) a mitochondrial preparation, (iii) a bacterial preparation and (iv) liver tissue. Some data for the factors discussed are given in Table 7.2.

Advantages of the liver tissue are shown in almost all areas. The slope, 50 mV per decade, is closer to the optimum Nernstian value of 59 mV per decade than any of the others. The detection limit is marginally better, 0.02 mM. The linear

Table 7.2 Response characteristics of glutamine biosensors [reproduced from Arnold and Rechnitz (1987)]

Parameter	Enzyme	Mitochondria	Bacteria	Tissue
Slope/mV per decade	33–41	53	49	50
Detection limit/M	6.0×10^{-5}	2.2×10^{-5}	5.6×10^{-5}	2.0×10^{-5}
Linear range/mM	0.15–3.3	0.11–5.5	0.1–10	0.064–5.2
Response time/min	4–5	6–7	5	5–7
Lifetime/days	1	10	20	30

range, 0.064–5.2 mM, is wider and the lifetime of the sensor is dramatically longer—30 days compared with 1 day for an enzyme sensor. The response time is marginally poorer, 5–7 min, compared with 4–5 min for the enzyme sensor.

7.6.2 Different Transducers

Table 7.3 compares the performance characteristics of a number of potentiometric-type biosensors.

Table 7.3 Performance characteristics of some potentiometric biosensors[a] [data taken from Kuan and Guilbault (1987)]

Type	Enzyme	Sensor[b]	Stability	Response time (min)	Range/M
Urea	Urease (25 U)	Cation (P)	3 weeks	0.5–1	10^{-2}–5×10^{-5}
Urea	Urease (75 U)	Cation (P)	4 months	1–2	10^{-2}–10^{-4}
Urea	Urease (100 U)	pH (P)	3 weeks	5–10	5×10^{-2}– 5×10^{-5}
Urea	Urease (10 U)	Gas (NH_3) (C)	4 months	2–4	5×10^{-2}– 5×10^{-5}
Urea	Urease (25 U)	Gas (CO_2) (P)	3 weeks	1–2	10^{-2}–10^{-4}
Glucose	GOD (100 U)	pH (S)	1 week	5–10	10^{-1}–10^{-3}
Glucose	GOD (10 U)	Iodide (C)	>1 month	2–8	10^{-3}–10^{-4}
L-Amino acids:					
General	L-AA oxidase				
		Cation (P)	2 weeks	1–2	10^{-2}–10^{-4}
		Iodide (C)	>1 month	1–3	10^{-3}–10^{-4}
L-Tyrosine	L-Tyrosine carboxylase	Gas (CO_2) (P)	3 weeks	1–2	10^{-1}–10^{-4}
L-Glutamine	Glutaminase	Cation (S)	2 days	1	10^{-1}–10^{-4}
L-Glutamic acid	Glutamate dehydrogenase	Cation (S)	2 days	1	10^{-1}–10^{-4}
L-Aspargine	Asparginase	Cation (P)	1 month	1	10^{-2}–5×10^{-5}
D-Amino acids					
(general)	D-AA oxidase	Cation (P)	1 month	1	10^{-2}–5×10^{-5}
Penicillin	Penicillinase				
	(400 U)	pH (P)	1–2 weeks	0.5–1	10^{-2}–10^{-4}
	(1000 U)	pH (S)	3 weeks	2	10^{-2}–10^{-4}
Amygdalin	β-Glucooxidase	Cyanide (P)	3 days	10–20	10^{-2}–10^{-5}
Nitrate	Nitrate reductase	Ammonium (S)	1 day	2–3	10^{-2}–10^{-4}

[a]Performance is also affected by the amount of enzyme used. This is, of course, a general effect but in the table is particularly noticable in the cases of urea, glucose and penicillin. The original table also shows some variation with the method of immobilisation.
[b](P) = physical entrapment in polyacrylamide gel.
(C) = covalent bonding with glutaraldehyde with albumin to polyacrylic acid (or acrylamide), followed by physical entrapment.
(S) = soluble.

7.6.2.1 UREA BIOSENSORS

The most extensive comparison is for the urea sensors, for which data are given for all the usual modes of operation: cation (NH_4^+), pH, gas (NH_3) and gas (CO_2).

The pH sensor needs the most enzyme and the response is linear only below 5×10^{-3} M. It also has the longest response time. The ammonia gas electrode is the best overall, being stable for 4 months in addition to having the widest response range, using very little enzyme and having a fairly short response time. The cation electrode has the best response time, with an excellent response range, but is stable for only 3 weeks.

A urea biosensor based on an ENFET, with urease in the sensing gate, used changes in pH to measure the urea. It was used in a differential mode so that the initial rate of voltage change was proportional to the logarithm of urea concentration over the range 1.3–16.7 mM urea. If the sensor was stored at 4°C between measurements it continued to work satisfactorily for at least 2 weeks. The response time was 2 min. Selectivity was tested against glucose, creatinine and albumin. No interfering response was observed.

Another urea biosensor was based on an IrPdMOS semiconductor which at pH > 8 is fairly selective for ammonia. The urease (40 U) was contained in a flow column (40 × 2 mm i.d. Eupergit). A linear response was obtained up to 40 μM ($\pm 2\%$) with a detection limit of 0.2 M. The response rate permitted 20 assays per hour. The urease column gave unchanged operation for 1 month.

7.6.2.2 GLUCOSE BIOSENSORS

Potentiometric glucose electrodes are not used much but are interesting for comparison with the more usual amperometric types. The pH mode has the best range but uses a considerable amount of enzyme and is stable for only 1 week. The iodide electrode allows a lower detection range, has a longer lifetime and uses less enzyme. For glucose determination in blood a working range of 2×10^{-3}–5×10^{-2} M is desirable. There have been so many variations of the glucose electrode that an overall comparison is difficult. However, we show in Table 7.4 the performance data for some commercially produced glucose biosensors, including the original Yellow Spring Instruments Model 23A. The lifetime data should be viewed with caution as lifetime may signify different things with different biosensors, as discussed in Section 7.5.

The earliest glucose biosensors monitored oxygen, but for blood analysis the blood oxygen interferes, so the biosensors to be used for clinical work were based on hydrogen peroxide measurement. Some mediated sensors are included in the table, in particular the Medisense Exactech sensor based on ferrocene. The advantages of the TTF–TCNQ-mediated and GDH–QQP

Table 7.4 Comparison of performance factors of some glucose biosensors

Type	System	Range	Response time/min	Lifetime
Analytical Instruments (Japan) Glucorecorder	GOD-O_2	0-5 mM (±2%)	0.5	
Radelkis (Hungary)	OP-G1-7113-S	1.7-2.0 mM (±5%)	1.5	8 months
Yellow Springs Instruments (USA) Model 23A	GOD-H_2O_2	1.0-45.0 mM (±2%)	1.5	300 samples
ZWG Academy of Sciences (Germany) Glukometer GKM 01	GOD-H_2O_2	0.5-50.0 mM (±1.5%)	0.7-1	1000 samples
Hofmann-LaRoche (Switzerland) Glucose Analyser 5410	GOD-$[Fe(CN)_6]^{3-}$	2.5-27.5 mM (±1.5%)	1	8 weeks
Inst. Techn. Chem. Acad. Sciences (Germany)	GOD-quinone	0-55.5 mM (±3%)	4	8 weeks
Medisense (UK) Exactech	GOD-ferrocene	1-30 mM (±1%)	0.5	>1 year
	GOD-TTF-TCNQ	0.5-20 mM		100 days
	GDH-QQP	1-70 mM	<0.3	8h
	Con A-fluorescent dextran (optrode)	2.0-25 mM (±0.5%)		15 days
	Hexokinase-bacterial luciferase-ATP	2-100 pmol		
	Hexokinase-O_2-thermistor	0.5-25 mM (±0.6%)	1.5	

sensors are not immediately apparent from the data. The glucose dehydrogenase–quinoprotein sensors produce a much larger current than GOD-based sensors and are not affected by ambient oxygen. The enzyme is effectively 'wired' directly to the electrode.

Table 7.3 also shows some applications of potentiometric biosensors to the analysis of various amino acids. They mostly involve ammonium cation electrodes. There are some surprising differences between the general sensor for the D- and L-forms, both using the appropriate amino acid oxidases and the same method of immobilisation (polyacrylamide gel entrapment). Stability is given as 1 month for D-forms but 2 weeks for L-forms. The amount of enzyme needed is 50 U (D), but 10 U (L). The concentration ranges are very similar, as are the response times (2 min). These biosensors will respond to D- and L-leucine, D- and L-methionine and D- and L-phenylalanine, and also work with L-cysteine, L-tyrosine and L-tryptophan and with D-alanine, D-valine, D-norleucine and D-isoleucine.

7.6.2.3 URIC ACID

A number of comparative studies have been made of biosensors for uric acid, a knowledge of which is important in haematology disorders. The normal range is 140–420 mol l^{-1}. Uric acid is oxidised in the presence of uricase by oxygen:

$$\text{uric acid} + O_2 \xrightarrow{\text{uricase}} \text{allantoin} + CO_2 + H_2O_2$$

As with glucose, several modes of operation are possible:
  direct measurement of oxygen—linear up to 0.5 mM ($\pm5\%$), operation for up to 100 days;
  measurement of hydrogen peroxide—linear up to 3.0 mM ($\pm2\%$) operational for 17 days (1000 samples);
  use of a mediator $[Fe(CN)_6]^{3-}$ via horseradish peroxidase—linear up to 0.035 mM and operational for 40 days.

7.7 SOME FACTORS AFFECTING THE PERFORMANCE OF BIOSENSORS

7.7.1 Enzyme Amount

Enzymes are catalysts, not consumed by the reaction, and the precise amount (or concentration) is not crucial for the operation of a biosensor. However, there are limiting factors. If we consider the Michaelis–Menten equation, the rate of reaction is directly proportional to the enzyme concentration:

$$v = \frac{k[E_0][S]}{K_m + [S]}$$

Provided that there is sufficient enzyme present so that this process is not rate limiting, there is no problem. However, if there is too much enzyme or if the quality of the enzyme preparation is poor, so that considerable material is needed to provide sufficient units of enzyme activity, the excess of material can affect the rates of mass transport (principally diffusion) to the transducer. This factor is seldom mentioned in papers about biosensors. An example is given in Table 7.3 for urea, showing that trebling of the amount of enzyme from 25 to 75 U of urease caused a dramatic improvement in lifetime from 3 weeks to 4 months, with a slight lengthening of the response time and a slight deterioration in the detection limit. However, in general, there is little systematic published information.

7.7.2 Immobilisation Method

This has been discussed in Chapter 3. In summary, chemical (covalent) methods result in longer lifetimes, but can limit the response, by blocking the mass transport processes, and *vice versa* for physical methods. However, the chemical method may sometimes damage the enzyme, causing a further diminution in response. This is counterbalanced by the more rapid loss of enzyme from a weakly bound physical method (entrapment or adsorption).

7.7.3 pH of Buffer

It is normally necessary to control the pH of the test solution fairly carefully. Commonly a phosphate buffer at pH 7.4 is used. However, the optimum pH is very much dependent on the electron transfer mediator being used. Wilson and Turner (1992) provided a survey for GOD, and showed that there are three groups of mediators depending on the pH:

(i)   pH optimum 5.6 (citrate buffer), such as quinones and oxygen;
(ii)   pH optimum 7.5 (phosphate buffer), such as diamimnes, ferrocenes and TTF–TCNQ;
(iii) pH < 4, such as $[Fe(CN)_6]^{3-}$, indophenols.

7.8 REFERENCES

M. A. Arnold and G. A. Rechnitz (1987) 'Biosensors based on plant and animal tissue', in A. P. F. Turner, I. Karube and G. S. Wilson (Editors), 'Biosensors: Fundamentals and Applications', Oxford University Press, Chap. 3, pp. 30–59.

T. D. Gibson and J. R. Woodward (1992) 'Proteins stabilisation in biosensor system', in P. G. Eldman and J. Wang (Eds), *Biosensors and Chemical Sensors*, American Chemical Society, Washington, D.C., Chap. 5, pp. 40–55.

T. D. Gibson, J. N. Hulbert, S. M. Parker, J. R. Woodward and I. J. Higgins (1992) 'Extended shelf life of enzyme-based biosensors using a novel stabilisation system', *Biosensors Bioelectron.* **7,** 701.

S. S. Kuan and G. G. Guilbault (1987) 'Ion selective electrodes and biosensors based on ISEs' in A. P. F. Turner, I. Karube and G. S. Wilson (Eds), *Biosensors: Fundamentals and Applications*, Oxford University Press, Oxford, Chap. 9, pp. 135–152.

K. R. Rogers and J. N. Lin (1992) 'Biosensors for environmental monitoring', *Biosensors Bioelectron.* **7,** 317.

F. W. Scheller, D. Pfeiffer, F. Schubert, R. Renneberg and D. Kirsten (1987) 'Application of enzyme-based amperometric biosensors to the analysis of "real" samples', in A. P. F. Turner, I. Karube and G. S. Wilson (Eds), *Biosensors: Fundamentals and Applications*, Oxford University Press, Oxford, Chap. 18, pp. 315–346.

S. G. Weber and A. Webers (1993), 'Biosensor calibration. *In situ* recalibration of competitive binding sensors', *Anal. Chem.*, **65,** 223.

R. Wilson and A. P. F. Turner (1992) 'Glucose oxidase: an ideal enzyme', *Biosensors Bioelectron.* **7,** 165.

Chapter 8

Important Applications

8.1 THE THREE GENERATIONS OF BIOSENSORS

Sometimes these three modes of oxidation are referred to as first-, second- and third-generation biosensors:

first generation—oxygen electrode-based sensors;
second generation—mediator-based sensors;
third generation—directly coupled enzyme–electrodes.

However, there is some evidence that the mode of action of conducting salt electrodes is really the same as that of a mediator, so that the third-generation description may not be strictly accurate.

8.1.1 First Generation—Oxygen Electrode (Clark, 1987; Hall, 1990)

The original glucose enzyme electrode used molecular oxygen as the oxidising agent:

$$\text{glucose} + O_2 \xrightarrow{\text{glucose oxidase}} \text{gluconic acid} + H_2O_2$$

The reaction is followed by measuring the decrease in oxygen concentration using a Clark oxygen electrode. This was first developed in 1953 and uses the voltammetric principle of electrochemically reducing the oxygen. The cell current is directly proportional to the oxygen concentration. The glucose oxidase is immobilised in polyacrylamide gel on a gas-permeable membrane covering the electrode, which consists of a platinum cathode and a silver anode. Figure 8.1 shows a typical glucose sensor of this type. This system, in addition to being one of great practical importance in the medical field, is also a useful model system on which many other biosensors can be based.

Many other biosensors can be developed which use oxidases and oxygen and a list of some of them is given in Table 8.1.

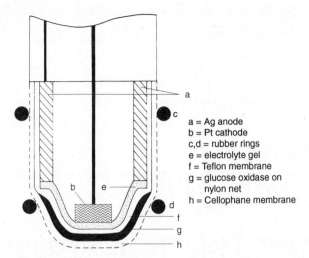

a = Ag anode
b = Pt cathode
c,d = rubber rings
e = electrolyte gel
f = Teflon membrane
g = glucose oxidase on
    nylon net
h = Cellophane membrane

Figure 8.1 Clark-type glucose electrode using two membranes. Reproduced by permission of the Open University Press from Hall (1990)

Table 8.1 Some oxidases used in biosensors

Analyte	Enzyme	Response time/min	Stability/days
Glucose	Glucose oxidase	2	>30
Cholesterol	Cholesterol oxidase	3	7
Monoamines	Monoamine oxidase	4	14
Oxalate	Oxalate	4	60
Lactate	Lactate oxidase		
Formaldehyde	Aldehyde oxidase		
Ethanol	Alcohol oxidase		
Glycollate	Glycollate oxidase		
NADH	NADH oxidase		

Reproduced by permission of the Open University Press from Hall (1990)

Although this device worked, it raised a number of problems. First, the ambient level of oxygen needed to be controlled and constant, otherwise the electrode response to the decrease in oxygen concentration would not be proportional to the decrease in glucose concentration. Another problem was that at the fairly high reduction potentials needed to reduce oxygen, $-0.7\,V$,

$$O_2 + e^- \longrightarrow O_2^-$$

other materials might interfere. The first way round this was to measure the oxidation of the hydrogen peroxide product:

$$H_2O_2 \longrightarrow 2H^+ + 2e^- + O_2$$

This could be done by setting the electrode potential to +0.65 V. This is still fairly high in the opposite sense, but now the problem could be interference from ascorbic acid, which is itself oxidised at this potential and is commonly present in biological samples.

A number of attempts have been made to regulate the oxygen level. Some are based on the fact that in the presence of the common enzyme catalase, hydrogen peroxide is decomposed to water and oxygen:

$$H_2O_2 \xrightarrow{\text{catalase}} H_2O + O_2$$

However, only half the required oxygen is produced and only then if all the hydrogen peroxide is recycled—in fact only about 50% can be recycled. An alternative is to reoxidise the water to oxygen at the anode:

$$H_2O - 2e^- \longrightarrow 2H^+ + O_2$$

Again, the standard electrode potential for this is very high at +1.23 V, and the application of such a potential would be likely to oxidise interferents. Some success has been obtained with an oxygen-stabilised electrode in which a separate oxygen generation circuit is used, controlled through a feedback amplifier from the analysing oxygen electrode. This is shown in Figure 8.2 (Enfors, 1987).

The operational amplifier compares the measured current due to the analysis of the oxygen with a standard potential. This is fed back to control

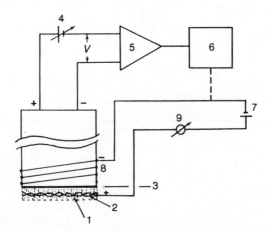

Figure 8.2 Generator to give a constant oxygen concentration. 1, Immobilised enzymes; 2, platinum net; 3, Teflon membrane of oxygen electrode; 4, reference voltage; 5, differential amplifier; 6, PID controller that controls the current through the electrolysis circuit to keep the differential voltage (V) zero; 7, voltage source of electrolysis circuit; 8, platinum coil around electrode; 9, microammeter. Reproduced by permission of Oxford University Press from Enfors (1987)

the electrolysis potential of the oxygen-generating circuit. The glucose oxidase is mixed with catalase and is embedded in a platinum gauze electrode that is the anode of generating circuit.

Another alternative is to control the oxygen level by the rate at which oxygen-containing buffer is pumped through the cell.

8.1.2 Second Generation—Mediators (Cardosi and Turner, 1987; Davis, 1987; Hall, 1990)

The idea was developed to replace oxygen with other oxidising agents—electron transfer agents—which were reversible, had appropriate oxidation potentials and whose concentrations could be controlled. Transition metal cations and their complexes were mostly used. These agents are usually called *mediators*. Many mediators are based on iron ions or its complexes:

$$Fe(III) + e^- \rightleftharpoons Fe(II)$$

Free iron(III) ions do not make good mediators as they are subject to hydrolysis and precipitation as iron(III) hydroxide, $Fe(OH)_3$.

A common complex which is sometimes used is hexacyanoferrate(III), $[Fe(CN)_6]^{3-}$, formerly known as ferricyanide. However, the most successful mediators have been the ferrocenes. Ferrocenes consist of a sandwich of an iron ion between two cyclopentadienyl anions, as shown in Figure 8.3.

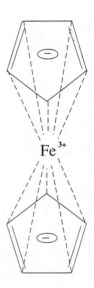

Figure 8.3 The structure of ferrocene

$$Fe^{3+}_{aq} + e^- \rightleftharpoons Fe^{2+}_{aq} \qquad \text{has } E^0 = +0.53 \text{ V}$$

$$\downarrow H_2O \qquad\qquad \downarrow H_2O$$

$$Fe(OH)_3 + 3H^+ \quad Fe(OH)_2 + 2H^+ \qquad \text{(hydrolysis)}$$

$$[Fe^{III}(CN)_6]^{3-} + e^- \rightleftharpoons [Fe^{II}(CN)_6]^{4-} \qquad \text{has } E^0 = +0.45 \text{ V}$$

$$[Fe^{III}(Cp)_2]^+ \ e^- \rightleftharpoons Fe^{II}(Cp)_2 \qquad \text{has } E^0 = +0.165 \text{ V}$$

$$\text{ferrocene} \qquad\qquad [E_p(Ox) = +0.193 \text{ V}$$

$$E_p(R) = +0.137 \text{ V}]$$

where Cp = cyclopentadienyl.

Taking the example of glucose, the operation of a mediator is as follows:

$$\text{glucose} + GOD_{Ox} \rightleftharpoons \text{gluconolactone} + GOD_{Red} + 2H^+$$

$$GOD_{Red} + 2Fc^+ \rightleftharpoons GOD_{Ox} + 2Fc$$

$$2Fc - 2e^- \rightleftharpoons 2Fc^+$$

where Fc = ferrocene.

The actual oxidation of the glucose is carried out by the FAD component of the glucose oxidase, which is converted into $FADH_2$. The $FADH_2$ is reoxidised to the FAD by the Fc^+ (mediator), then the Fc is reoxidised to Fc^+ directly at an electrode. The current flowing through the electrode is an amperometric measure of the glucose concentration. This is better shown in the cyclic diagram in Figure 8.4.

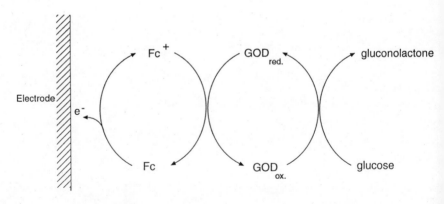

Fc = ferrocene

GOD = glucose oxidase

Figure 8.4 A ferrocene-mediated biosensor for glucose

The properties of a good mediator are the following:

(i)   It should react rapidly with the enzyme.
(ii) It should show reversible (i.e. fast) electron transfer kinetics.
(iii) It should have a low over-potential for regeneration.
(iv) It should be independent of pH.
(v)   It should be stable in both the oxidised and reduced forms.
(vi) It should not react with oxygen.
(vii) It should be non-toxic.

Ferrocenes fit all these criteria.

As observed previously, an oxygen electrode is operated at -0.6 V, at which potential it is also likely to reduce ascorbic acid, which is normally present in most enzyme or cell preparations in large amounts.

Table 8.2 list some important redox potentials.

Table 8.2 Redox potentials of some important reactions at pH 7

Reaction	E/V vs NHE	Reaction	E/V vs NHE
Acetate–acetaldehyde	-0.60	Oxaloacetate–L-malate	-0.17
Acetone–propan-2-ol	-0.43	Ubiquinone–reduced ubiqinone	0.00
H^+–H_2	-0.42	Fumarate–succinate	$+0.03$
Xanthine–hypoxanthine	-0.37	Dehydroascorbate–ascorbate	$+0.06$
NAD^+–NADH	-0.32	Ferrocene	$+0.165$
Oxidised–reduced glutathione	-0.23	O_2–H_2O_2	$+0.31$
Cystine–cysteine	-0.22	$[Fe(CN)_6]^{3-}$–$[Fe(CN)_6]^{4-}$	$+0.45$
Acetaldehyde–ethanol	-0.20	Fe^{3+}–Fe^{2+}	$+0.53$
Pyruvate–L-malate	-0.19	O_2–H_2	$+0.82$

The ring(s) of the cyclopentadienyl group may have various substituent groups attached. The presence of these groups affects the properties of the ferrocene, particularly the redox potential, but also the rate constant for electron transfer to the enzyme. Examples are shown in Table 8.3.

The solubility is also affected, which is important in formulating the biosensor. Thus, 1,1′-dimethylferrocene is insoluble in water and has an $E^\ominus$ of $+0.1$ V and a rate constant for reaction with glucose oxidase of 0.8×10^{-5} dm^3 mol^{-1} s^{-1}, whereas ferrocene monocarboxylic acid is fairly soluble in water and has an $E^\ominus$ of $+0.275$ V and a rate constant of 2.0×10^{-5} dm^3 mol^{-1} s^{-1}.

Many other suitable mediators are available and can be classified into 'natural' and 'artificial' electron mediators. The former include molecules

Table 8.3 Redox potentials of substituted ferrocenes and electron transfer rate constants for glucose oxidase oxidation by ferricinium derivatives

Ferrocene derivative	E/V vs SCE	$k_s/10^5\,\mathrm{dm^3\,mol^{-1}\,s^{-1}}$
1,1'-Dimethyl	0.100	0.8
Acetic acid	0.142	
Ferrocene	0.165	0.3
Amidopentylamidopyrrole	0.200	2.07
Aminopropylpyrrole	0.215	0.75
Vinyl	0.253	0.3
Monocarboxylic acid	0.273	2.0
1,1'-Dicarboxylic acid	0.290	0.3
Methyltrimethylamino	0.387	5.3
Polyvinyl	0.435	

such as the cytochromes, ubiquinone, flavoproteins and ferridoxins. Other artificial mediators include many dyestuffs such as methylene blue, phthalocyanines and viologens. Table 8.4 gives a comparison of the redox potentials of some of these mediators. The structures of some mediators are shown in Figure 8.5.

(a)   Redox proteins

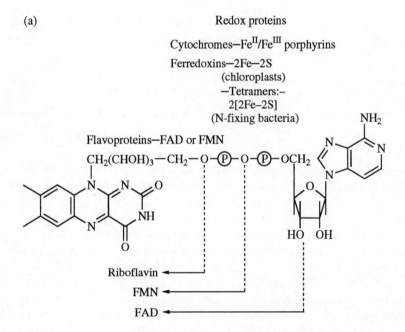

Cytochromes—Fe^{II}/Fe^{III} porphyrins

Ferredoxins—2Fe—2S
(chloroplasts)
—Tetramers:–
2[2Fe–2S]
(N-fixing bacteria)

Flavoproteins—FAD or FMN

(b) Other mediators

Viologens

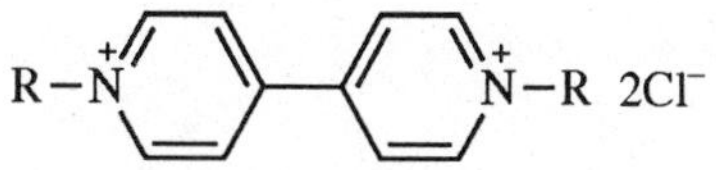

$$R-\overset{+}{N}\text{⟨⟩}\text{⟨⟩}\overset{+}{N}-R \quad 2Cl^-$$

$R = Me$ or $C_6H_5CH_2-$

Methylene blue

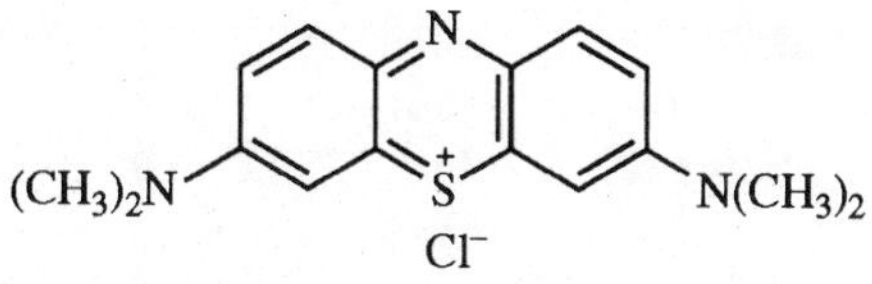

$(CH_3)_2N$ ... $\overset{+}{S}$ ... $N(CH_3)_2$
Cl^-

TMPD (Tetramethylphenylenediamine)

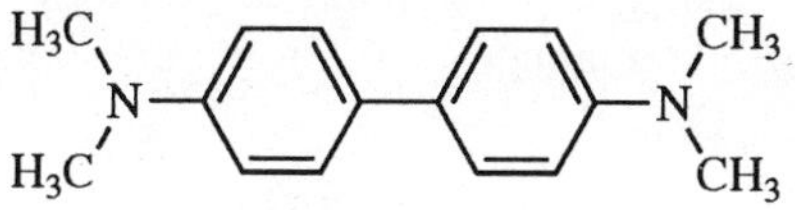

H_3C , H_3C $-N-$ ⟨⟩ ⟨⟩ $-N-$ CH_3 , CH_3

Phenazine methosulphate

$CH_3 \quad HSO_4^-$
$\overset{+}{N}$
N

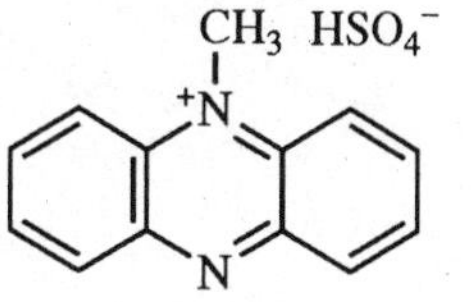

Figure 8.5 Some mediators: (a) 'natural' mediators; (b) 'artificial' mediators

Table 8.4 'Natural' and 'artificial' mediators and their redox potentials at pH 7

'Natural' mediators	E/V vs NHE	'Artificial'	E/V vs NHE
Cytochrome a_3	+0.29	Hexacyanoferrate(III)	+0.45
Cytochrome c_3	+0.24	2,6-Dichlorophenol	+0.24
Ubiquinone	+0.10	Indophenol	+0.24
Cytochrome b	+0.08	Ferrocene	+0.17
Vitamin K_2	−0.03	Phenazine methosulphate	+0.07
Rubredoxin	−0.05	Methylene blue	+0.04
Flavoproteins	−0.4 to +0.2		
		Phthalocyanine	−0.02
FAD–FADH$_2$	−0.23	Phenosafranine	−0.23
FMN–FMNH$_2$	−0.23	Benzylviologen	−0.36
NAD$^+$–NADH	−0.32	Methylviologen	−0.46
NAPD$^+$–NADPH	−0.32		
Ferrodoxin	−0.43		

8.1.2.1   RATE CONSTANTS

In general, we can write the rate mechanism as follows;

$$Red \rightleftharpoons Ox + e^-$$

$$E_{Red} + Ox \xrightarrow{k_1} E_{Ox} + R$$

$$E_{Ox} + glucose \xrightarrow{k_2} E_{Red} + gluconolactone$$

If $k_1 < 10\,k_2/[glucose]$, then k_2 is fast and k_1 is the rate-determining step.

We can study the effect of mediators by cyclic voltammetry, (Rusling and Ito, 1991) and obtain an estimate of the rate constant. If we measure the cyclic voltammogram of a solution containing ferrocene moncarboxylic acid in a phosphate buffer (pH 7) also containing glucose, we obtain the typical reversible shape of the ferrocene cyclic voltammogram, as shown in Figure 8.6 (A). If we then add glucose oxidase to this solution we obtain the catalyic wave with a greatly enhanced oxidation peak and no reduction peak, as shown in Figure 8.6 (B).

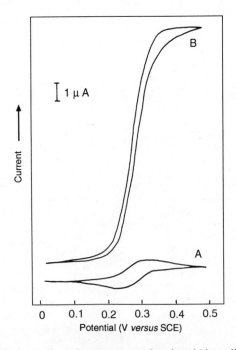

Figure 8.6   A catalytic cyclic voltammogram showing (A) cyclic voltammogram of ferrocene monocarboxylic acid in the presence of glucose and (B) as for (A) but with the addition of glucose oxidase. Reproduced by permission of Analytical Chemistry from Cass *et al.* (1984)

The rate constant is proportional to the relative height of the catalytic wave, so that

$$i_k/i_d = f(\log k_1/v)$$

The relationship is shown in a graph of i_k/i_d versus $(k_f/a)^{1/2}$ (where $a = nFv/RT$); i_k = catalytic current (with GOD); i_d = diffusion controlled current in Figure 8.7.

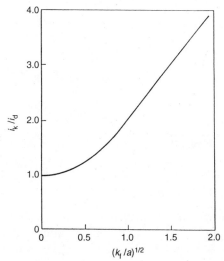

Figure 8.7 Theoretical plot of the ratio of the kinetic-to diffusion-controlled peak current, i_k/i_d, versus the kinetic parameter $(k_f/a)^{1/2}$. Reproduced by permission of Oxford University Press from Davis (1987)

8.1.2.2 FORMATION OF BIOSENSORS USING MEDIATORS

There are many ways in which mediators can be incorporated into biosensors. In the experiment described above the components are all in solution. In a biosensor both the enzyme and the ferrocene must be immobilised on the electrode.

The simplest approach is to mix the mediator with carbon paste (liquid paraffin mixed with graphite powder) in a carbon paste electrode, then the enzyme is adsorbed on the surface and held in place with a membrane (Hall 1990). This is shown in Figure 8.8.

A more sophisticated approach was used by Cass *et al.* (1984). Graphite foil with the edge plane exposed was coated with dimethylferrocene by evaporation from a toluene solution. Glucose oxidase in a buffer was then immobilised on the surface by reaction with 1-cyclohexyl-3-(2-morpholinoethyl)carbodiimide-*p*-methyltoluenesulphonate. The sensor was then covered with a Nuclepore membrane.

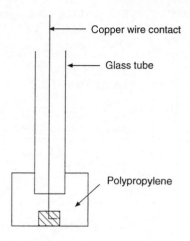

Copper wire contact

Glass tube

Polypropylene

Figure 8.8   A carbon paste electrode

8.1.3   Third Generation—Direct Enzyme–Electrode Coupling (Albery and Cranston, 1987; Bartlett, 1987)

It may seem strange that a mediator is needed to couple an enzyme to an electrode. Why would it not be possible to reduce (oxidase) an enzyme directly at an electrode? The problem is that proteins tend to be denatured on electrode surfaces. Also, the electron transfer may be slow and irreversible and hence require an excessively high overpotential.

A possible approach is to modify the surface, e.g. with 4,4'-bipyridyl on a gold electrode. The bipyridyl is not itself electroactive, nor is it a mediator. It forms weak hydrogen bonds with lysine residues on the enzyme. The binding is temporary.

A better solution was developed by Albery and Cranston (1987) and Bartlett (1987) using organic conducting salt electrodes. Tetrathiafulvalene (TTF) is reversibly oxidised, and tetracyanoquinodimethane (TCNQ) (Figure 8.9) is

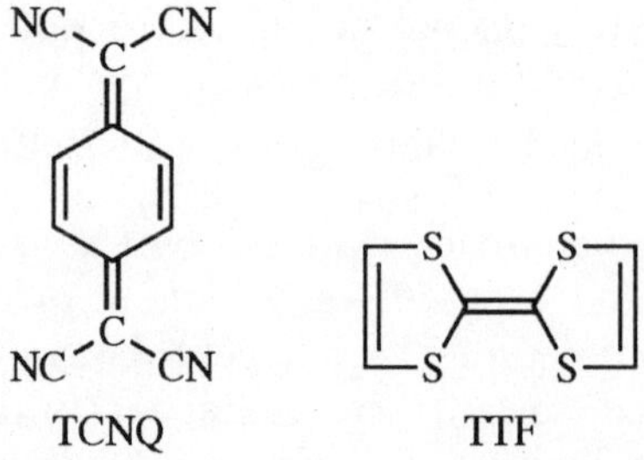

Figure 8.9   Tetracyanoquinodimethane (TCNQ) and tetrathiafulvalene (TTF)

similarly reversibly reduced. A pair of these molecules form a charge-transfer complex. It is found that when these are incorporated into an electrode the surface is highly reversible and stable to many enzymes. Another important molecule is *N*-methylphenothiazine (NMP), which is sometime preferred to TTF.

These conducting salts can be built into electrodes in three ways: as single crystals, as pressed pellets or as a paste with graphite powder. They vary slightly in their properties, in that the higher the crystalline form, the better is the reversibility, but then, of course, the construction technique is more difficult.

8.1.3.1 DIRECT ENZYME–ELECTRODE COUPLING (YE *ET AL.*, 1993)

Recently, immobilisation techniques have been developed to 'wire' an enzyme directly to an electrode, facilitating rapid electron transfer and hence high current densities. In general they involve an *in situ* polymerisation process using a redox polymer. An example of this method used a glucose dehyrogenase (GDH) containing the redox centre pyrroloquinolinequinone (PQQ), which was 'wired' to the glassy carbon electrode through a redox polymer, polyvinylpyridine partially N-complexed with [osmium bis(bipyridine) chloride]$^{2+}$ and quaternised with bromoethylamine (POs-EA), cross-linked with poly(ethylene glycol 400 diglycidyl ether) (PEGDE). This oxygen-insensitive biosensor produced a very high current density of 1.8 mA cm^{-2} with 70 mM glucose, said to be three times higher than with a GOD sensor. The dissolved enzyme had a half-life of 5 days, but in continuous operation the current had decayed to the baseline in 8 h.

8.2 NADH/NAD$^+$ (Albery and Cranston, 1987; Hall, 1990)

Nicotinamide adenine dinucleotide is a very common cofactor in many biochemical processes, coupling a hydrogen transfer reaction with an enzyme reaction:

$$NAD^+ + RR'CHOH \rightarrow NADH + RR'C{=}O + H^+$$

The structure of NADH is shown in Figure 8.10a and its redox mode in Figure 8.10b.

Unfortunately the redox behaviour of NAD$^+$–NADH is rather irreversible at an electrode. Electrochemical reduction of NAD$^+$ does not give NADH but goes to a dimer. NADH can be oxidised electrochemically to NAD$^+$, but at a substantial overpotential—above the standard redox potential. One can use a modified electrode, i.e. one coated with a suitable surface mediator, but such electrodes lack long-term stability. The use of

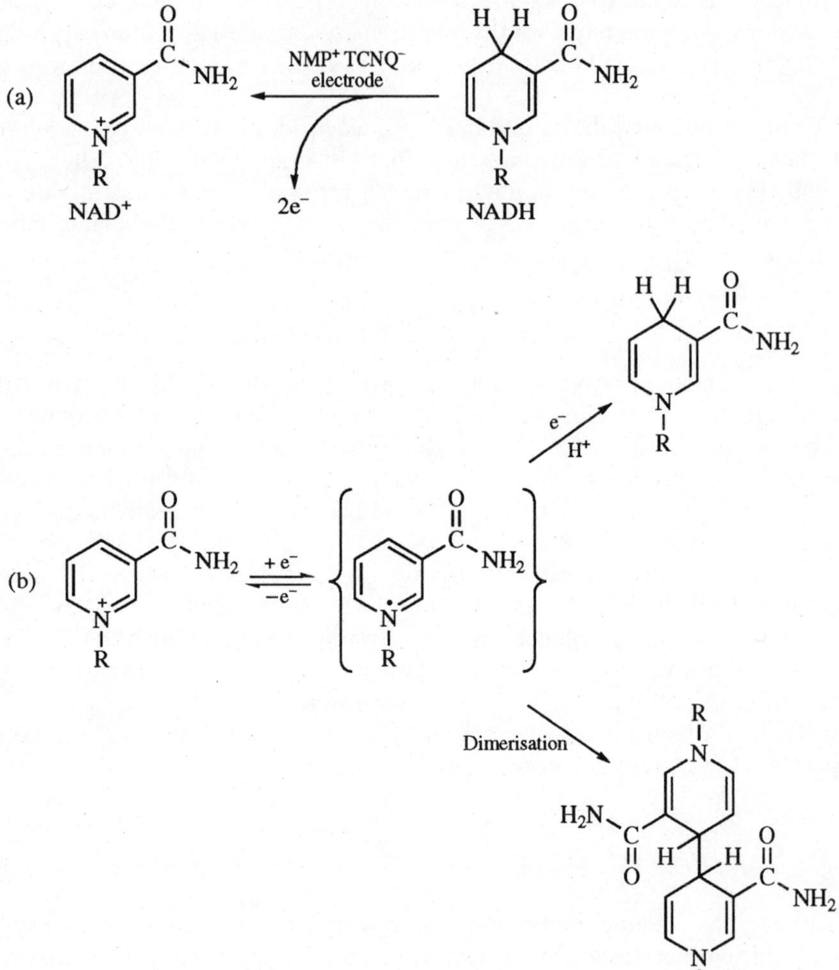

Figure 8.10 Structures of (a) nicotine adenine dinucleotide (NAD) and (b) reduced NAD (NADH), and their reduction, oxidation and dimerisation. Reproduced by permission of the Open University Press from Hall (1990)

conducting salt electrodes overcomes this difficulty. NADH can be oxidised at $-0.2\,V$ (vs Ag/AgCl) on an NMP$^+$ TCNQ$^-$ electrode. It can be used in a biosensor for lactate with lactate dehydrogenase (LDH). The NADH was recycled on a glassy carbon electrode at $+0.75\,V$, considerably above the standard potential of $-0.32\,V$.

$$CH_3CHOHCO_2^- + NAD^+ \xrightarrow{LDH} CH_3COCO_2^- + NADH + H^+$$

$$NADH \longrightarrow NAD^+ + e^- + H^+ \qquad (GC +0.75\,V)$$

$$(NMP^+ TCNQ^- -0.2\,V)$$

There is a very large number of useful reactions that can be driven by such systems, particularly those involving dehyrogenases. In 1987 this number was estimated to be >250.

One problem which has not been overcome is that NADH (and NAD$^+$) are expensive and not very stable, nor can they easily be immobilised on a biosensor. Therefore, they have to be added to the analysis solution.

In addition to the lactate–pyruvate example mentioned above, ethanol can be readily measured by a biosensor based on this technique:

$$C_2H_5OH + NAD^+ \xrightarrow{ADH} CH_3CHO + NADH + H^+$$

Another important bioassay is for cholesterol:

$$cholesterol + NAD^+ \xrightarrow{ChDH} cholestenone + NADH + H^+$$

Other analytes that can be assayed in this way include L-amino acids, glycolic acid and, of course, NADH itself. While biosensors can be constructed for the analytes using conducting salt electrodes, a ferrocene-mediated electrode can be used if the ferrocene is coupled to the NADH via the enzyme diaphorase (lipoamide dehydrogenase) (Figure 8.11).

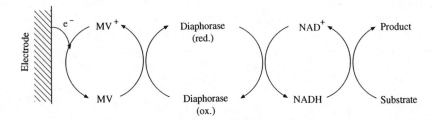

Figure 8.11 Reversible redox behaviour of NAD–NADH linked by diaphorase to a methyl viologen mediator

Table 8.5   Some NADH coupled assays

$$\text{pyruvate} + \text{NADH} + \text{H}^+ \xrightarrow{\text{LDH}} \text{lactate} + \text{NAD}^+$$

$$\text{oxalacetate} + \text{NADH} + \text{H}^+ \xrightarrow{\text{MDH}} \text{malate} + \text{NAD}^+$$

$$\text{EtOH} + \text{NAD}^+ \xrightarrow{\text{ADH}} \text{CH}_3\text{CHO} + \text{NADH} + \text{H}^+$$

$$\text{G-6-P} + \text{NAD}^+ \xrightarrow{\text{G-6-PDH}} \text{glucono-6'-lactone-6-P} + \text{NADH} + \text{H}^+$$

$$\text{CH}_4 + \text{NAD}^+ + \text{H}_2\text{O} \xrightarrow{\text{MMO}} \text{CH}_3\text{OH} + \text{NADH} + \text{H}^+$$

$$\text{HCO}_2\text{H} + \text{NAD}^+ \longrightarrow \text{CO}_2 + \text{NADH} + \text{H}^+$$

$$\text{NO}_3^- + \text{NADH} + \text{H}^+ \xrightarrow{\text{N(ate)R}} \text{NO}_2^- + \text{H}_2\text{O} + \text{NAD}^+$$

$$\text{NO}_2^- + 3\text{NADH} + 4\text{H}^+ \xrightarrow{\text{N(ite)R}} \text{NH}_3 + 2\text{H}_2\text{O} + 3\text{NAD}^+$$

$$2\text{Fe(CN)}_6^{3-} + \text{NADH} \xrightarrow{\text{diaphorase}} 2\text{Fe(CN)}_6^{4-} + \text{NAD}^+ + \text{H}^+$$

$$\text{MV}^+ + \text{NADH} + \xrightarrow{\text{diaphorase}} \text{MV} + \text{NAD}^+$$

$$\text{androsterone} + \text{NAD}^+ \xrightarrow{\text{HSDH}} \text{5-androstane-3,17-dione} + \text{NADH}$$

$$\text{cholesterol} + \text{NAD}^+ \xrightarrow{\text{ChDH}} \text{cholestenone} + \text{NADH} + \text{H}^+$$

8.3 AMPEROMETRIC TRANSDUCERS

8.3.1 Glucose (Scheller *et al.*, 1987; Hall, 1990)

It is said that about half the research papers published on biosensors are concerned with glucose. In addition to its metabolic and medical importance, it makes a good standard compound on which to try out possible new biosensor techniques. There are in fact a number of different ways of determining glucose just using electrochemical transducers. Reference to this has already been made under other tranducer modes in Chapters 5 and 6.

The diagram in Figure 8.12 show the overall pattern. One can see that there are several different ways in which glucose may be determined. Glucose oxidase lies at the centre of them all.

(i) Because gluconic acid is formed as a product, there is a change in pH and so pH measurement can be used to monitor the reaction.

$$\text{glucose} + \text{O}_2 \xrightarrow{\text{GOD}} \text{gluconic acid} + \text{H}_2\text{O}_2$$

(ii) The reoxidation of the reduced form of GOD directly at the electrode is

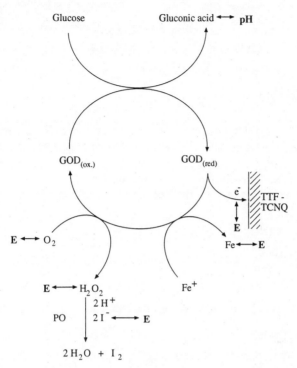

Figure 8.12 Diagram showing different pathways for the oxidation of glucose which can be followed by electrochemical sensors. Adapted by permission of the Open University Press from Hall (1990)

possible at special electrodes as described in Section 8.1.3.

glucose + GOD_{Ox} $\longrightarrow$ gluconic acid + GOD_{Red}

GOD_{Red} $-2e^-$ $\longrightarrow$ GOD_{Ox}

(iii) As oxygen is consumed in the reaction, the decrease in oxygen concentration can be monitored with a Clark oxygen electrode, as described in Section 8.1.1.

glucose + O_2 $\longrightarrow$ gluconic acid + H_2O_2

$O_2 + 2e^- +2H^+$ $\longrightarrow$ H_2O_2

(iv) An alternative is to monitor the hydrogen peroxide produced by the reduction of oxygen. This may be done directly by electrooxidation at +0.6 V.

$H_2O_2 - 2e^-$ $\longrightarrow$ $O_2 + 2H^+$

(v) The hydrogen peroxide can be used to oxidise iodide to iodine in the presence of peroxidase (PO) and the decrease in iodide concentration measured with an iodide-selective electrode.

$$H_2O_2 + 2HI \xrightarrow{PO} H_2O + I_2$$

(vi) The oxygen may be replaced by a mediator, such as ferrocene (see Section 8.1.2) which can be detected by electrochemical oxidation.

$$glucose + 2Fc^+ \xrightarrow{GOD} gluconic\ acid + 2Fc$$

8.3.2   Lactate (Scheller *et al.* 1987)

Lactate ($CH_3CHOHCO_2H$) is an important analyte because of its involvement in muscle action, following which its concentration in blood rises. There are four different enzymes which can be used, two being mediator driven and the other two oxygen driven. The processes in three cases leads to pyruvate (CH_3COCO_2H) and in the other to acetate.

(i) $lactate + NAD^+ \xrightarrow{LDH} pyruvate + NADH + H^+$

At the electrode

$$NADH \longrightarrow NAD^+ + 2e^- + H^+$$

(ii) $lactate + 2[Fe(CN)_6]^{3-} \xrightarrow{Cyt\,b_2} pyruvate + 2H^+ + 2[Fe(CN)_6]^{4-}$

At the electrode:

$$Fe^{2+} \longrightarrow Fe^{3+} + e^-$$

(iii) $lactate + O_2 \xrightarrow{LOD} pyruvate + H_2O_2$

At the electrode: H_2O_2 or O_2

(iv) $lactate + O_2 \xrightarrow{LMO} acetate + CO_2 + H_2O$

At the electrode: O_2

where LDH = lactate dehydrogenase, LOD = lactate oxidase, LMO = lactate monooxidase and Cyt b_2 = cytochrome b_2.

8.3.3   Cholesterol (Scheller *et al.* 1987)

Too much cholesterol in the body is thought to be associated with heart disease, so monitoring the level in the blood is becoming a routine health check analysis. The existing procedure is cumbersome and so a biosensor method would be very useful. At the time of writing, no successful commercialisation has been achieved. However, a number of successful procedures have been developed in the laboratory.

Cholesterol is the most basic steroid alcohol. It is the major constituent of gallstones, from which it can be extracted. Cholesterol commonly occurs in the bloodstream as an ester. Therefore, any method of analysis involving the hydroxy group needs a preliminary hydrolysis. This can be facilitated with the enzyme cholesterol esterase, which can be combined with the oxidising enzyme cholesterol oxidase, which catalyses oxidation to cholestenone.

Hall (1990) described three approaches, all involving ferrocene as a mediator. The first involves coupling through NAD⁺–NADH, diaphroase and then ferrocene to an electrode, as shown in Figure 8.13.

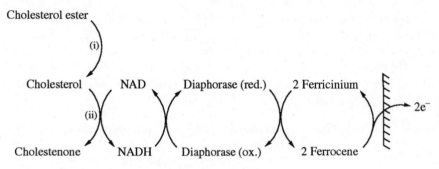

(i) cholesterol esterase; (ii) cholesterol dehydrogenase

Figure 8.13 A cholesterol biosensor using NAD

The second uses oxygen, which is converted into hydrogen peroxide, which is then coupled via perioxidase and ferrocene as shown in Figure 8.14.

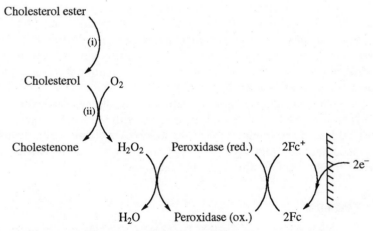

(i) cholesterol esterase; (ii) cholesterol oxidase

Figure 8.14 A cholesterol biosensor using peroxidase

The third directly couples cholesterol oxidase to ferrocene.

A novel method developed by Cassidy *et al.* (1993) uses a thin-layer cell containing $[Fe(CN)_6]^{3-}$ which shuttles oxygen backwards and forwards across the cell, setting up a steady-state situation. Such arrangements are sometimes called 'fuel cells' in biosensor publications.

8.3.4 Phosphate (Scheller *et al.*, 1987)

This is an interesting indirect assay using a glucose sensor. Glucose-6-phosphate is hydrolysed with acid phosphatase (AP) to free phosphoric acid (P_i) and glucose, which can be determined with a glucose biosensor. However, phosphate (P_i) inhibits the action of the phosphatase. The procedure then is to allow glucose-6-phosphate to react with phosphatase in the presence of the phosphate to be determined. This inhibits the reaction and so a reduced amount of glucose is formed.

$$\text{glucose-6-phosphate} \xrightarrow{\text{AP}} \text{glucose} + \text{phosphoric acid}$$

$$\text{glucose} + O_2 \xrightarrow{\text{GOD}} \text{gluconic acid}$$

8.3.5 Starch (Scheller *et al.*, 1987)

Starch is broken down by α-amylase to dextrins and maltose. Glucoamylase will break down the maltose to glucose, which can then be determined with a glucose biosensor. The usual method in this example is to measure the hydrogen peroxide produced by the oxygen–glucose oxidase reaction using the glucose at an electrode. However, there will be glucose in the original solution and this must be filtered out. A double-membrane filter is used. Everything can pass through the first membrane. Inside this membrane is glucose oxidase, oxygen and catalase. The interfering glucose is broken down to hydrogen peroxide and gluconolactone. The hydrogen peroxide is then oxidised by the catalase to oxygen. Hence no free glucose or hydrogen peroxide passes through the second membrane to the electrode, but maltose and oxygen do. Inside the second membrane is the glucoamylase and more glucose oxidase. These convert the maltose successively into glucose and then into gluconolactone and hydrogen peroxide, which is measured at the electrode. This scheme is illustrated in Figure 8.15.

8.3.6 Ethanol

This is an important target analyte because of the need to monitor blood-alcohol levels. Several methods have been developed.

One approach uses a microbial sensor using either *Acetobacter xylinium* or

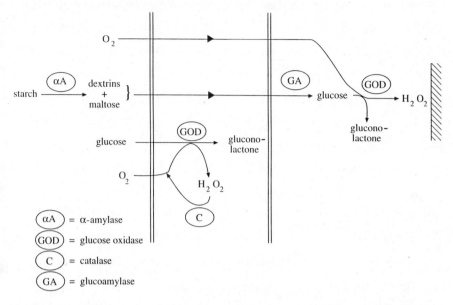

Figure 8.15 Principle of starch determination using a glucose-eliminating multi-layer sensor. Reproduced by permission of Oxford University Press from Scheller *et al.* (1987)

Trichosporon brassicae. These both catalyse the aerial oxidation of ethanol to acetic acid:

$$\text{ethanol} + O_2 \xrightarrow{\textit{A. xylinium}} \text{acetic acid} + H_2O$$

This reaction is followed with an oxygen electrode. The method has been developed commercially in Japan, using membrane encapsulation. It works over the range 5–72 mM.

An alternative mediated method uses alcohol dehydrogenase (ADH) coupled via NAD^+–NADH and $[Fe(CN)_6]^{3-/4-}$ to an electrode as shown in Figure 8.16.

There is also a bioluminescent detection method, also based on ADH and NAD^+ (see Chapter 6).

8.3.7 Carbon Monoxide (Turner *et al.*, 1984)

Despite the disappearance of coal gas, poisoning and death by carbon monoxide from motor car exhaust fumes, domestic solid fuel heater fumes or oil heating fumes are unfortunately all too common. Existing devices for detection of CO are expensive and not very selective. They are usually based

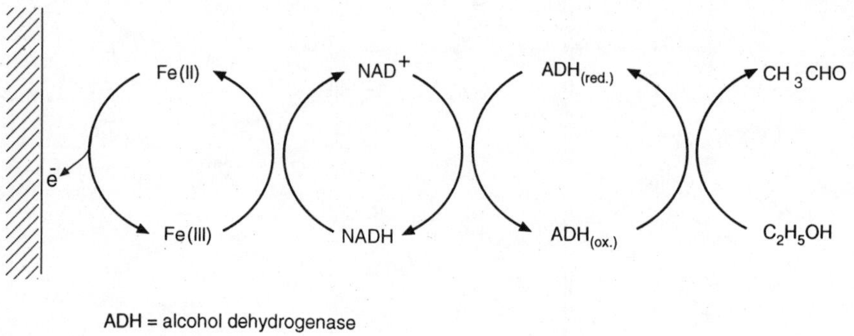

ADH = alcohol dehydrogenase

Figure 8.16   A biosensor for ethanol using alcohol dehydrogenase and NAD with an Fe(II)/(III) mediator

on infrared spectroscopy. The problem is that CO is such as very simple molecule and it behaves very similarly to oxygen in many situations. However, some bacteria have been found, especially in anaerobic cultures, which contain enzymes which will catalyse the oxidation of CO to CO_2.

8.3.8   Aspirin (Hilditch and Green, 1991)

Levels above about 3 mM aspirin in blood are toxic. The therapeutic dose is 1.1–2.2 mM. In blood, acetylsalicylic acid (aspirin) is converted into salicylic acid by hepatic esterases. The conventional method for the determination salicylic acid is spectrophoto-metric measurement of the complex formed with iron(III), which lacks specificity. An enzyme of bacterial origin, salicylate hydroxylase, catalyses the oxidation of salicylate to catechol via NADH, as shown in Figure 8.17. The catechol can be monitored by electrooxidation using a screen-printed carbon electrode. However, this method has technical problems, which are being addressed in our laboratories (Jones, Marchant and Zhou, 1994; Banat *et al.* (1994)).

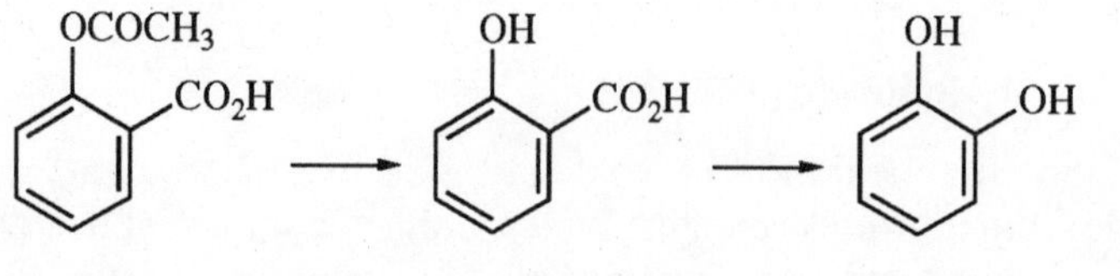

Figure 8.17   Enzyme-catalysed hydrolysis and breakdown of aspirin

8.3.9 Paracetamol (*N*-Acetyl-*p*-aminophenol) (Hilditch and Green, 1991)

Poisoning by excess paracetamol, nowadays the commonest component of analgesic tablets, can cause irreversible liver damage. It is therefore vital that its identity and concentration in people who have taken overdoses be discovered as soon as possible. A successful biosensor would be a very suitable device for doing this. Paracetamol can be readily oxidised at a carbon paste electrode, but this, of course is non-selective. However, the enzyme aryl-acylamidase will catalyse the hydrolysis of paracetamol to *p*-aminophenol, which is then electrochemically oxidisable at a much lower potential to quinoneimine, again using a diposable screen-printed carbon strip electrode. This shown in Figure 8.18.

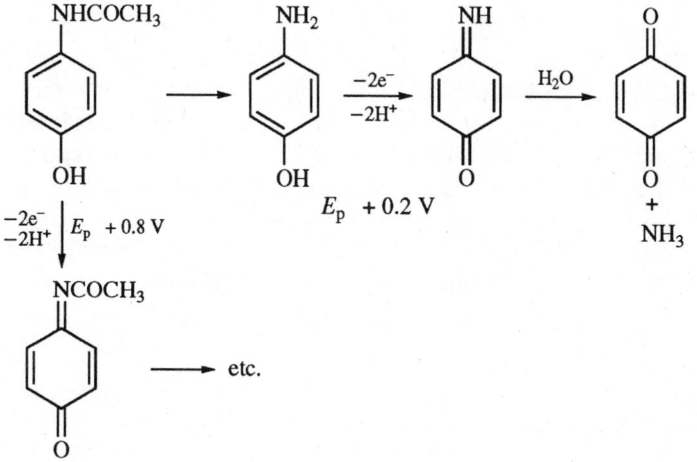

Figure 8.18 Enzyme-catalysed hydrolysis and oxidation of paracetamol compared with direct oxidation

8.4 POTENTIOMETIC BIOSENSORS (Kuan and Guilbault, 1987; Hall, 1990)

These are largely based on certain ion-selective electrodes or gas-selective electrodes.

8.4.1. pH-Linked

8.4.1.1 PENICILLIN

$$\text{penicillin} \xrightarrow{\text{penicillinase}} \text{penicilloate} + H^+$$

With this reaction the pH decreases and the response time is <30 s. The response slope is 52 mV per decade over the range 5×10^{-2}–10^{-4} M.

8.4.1.2 GLUCOSE

See Section 8.3.1.

8.4.1.3 UREA

Urea is the most important and first analyte to be determined with a potentiometric biosensor. It is hydrolysed with the aid of the common enzyme urease, which is present in jack bean meal:

$$CO(NH_2)_2 + H_2 \xrightarrow{\text{urease}} 2NH_4^+ + CO_3^{2-}$$

The analysis may be carried out in a number of ways. One can use a cation ISE for ammonia, one can make the solution alkaline and determine the liberated ammonia with an ammonia gas ISE or one can make the solution acid and determine the liberated carbon dioxide with a carbon dioxide gas ISE. With the aid of a suitable buffer such as histidine, one can also measure the reaction with a pH meter. Table 8.6 gives some comparative data on performance criteria for these various modes of analysis (see Chapter 7).

Table 8.6 Performance characteristics of some urea potentiometric biosensors [data taken from Kuan and Guilbault (1987)]

Enzyme	Sensor[a]	Stability	Response time/min	Range/M
Urease (25 U)	Cation (P)	3 weeks	0.5–1	10^{-2}–5×10^{-5}
Urease (75 U)	Cation (P)	4 months	1–2	10^{-2}–10^{-4}
Urease (100 U)	pH (P)	3 weeks	5–10	5×10^{-2}–5×10^{-5}
Urease (10 U)	Gas (NH_3) (C)	4 months	2–4	5×10^{-5}–5×10^{-5}
Urease (25 U)	Gas CO_2 (P)	3 weeks	1–2	10^{-2}–10^{-4}

[a](P) = physical entrapment in polyacrylamide gel; (C) = covalent bonding with glutarldehyde with albumin to polyacrylic acid (or acrylamide) followed by physical entrapment.

The most successful has been the NH_3 electrode with the urease attached to the polypropylene membrane of the NH_3 ISE. It has the highest selectivity and the lowest detection limit (10^{-6} M). It can achieve 20 assays per hour with a relative standard deviation of ±2.5% over the range 5×10^{-5}–10^{-2} M.

8.4.2 Ammonia-Linked

Various reactions can be utilised, as indicated below (for urea, see Section 8.4.1.3).

(i) Creatinine

$$\text{creatinine} \xrightarrow{\text{creatinase}} NH_3 + \text{creatine}$$

With the creatinase immobilised on the polypropylene membrane of an ammonia electrode, the electrode was stable for 8 months and 200 assays, and had a detection limit of 8×10^{-6} M.

(ii) L-Phenylalanine

$$\text{L-phenylalanine} \xrightarrow{\text{phenylalanine ammonia-lyase}} NH_3 + \textit{trans}\text{-cinnamate}$$

This sensor is very highly selective, but has a poor range and slow response.

(iii) Adenosine

$$\text{adenosine} \xrightarrow{\text{adenine deaminase}} NH_3 + \text{inosine}$$

The adenine deaminase is cross-linked with glutaraldehyde on the ammonia electrode.

(iv) Aspartame

$$\text{aspartame} \xrightarrow{\text{L-aspartase}} NH_3 + C_6H_5CH_2CH(CO_2H)NHCOCH{=}CHCO_2H$$

8.4.3 Carbon Dioxide-linked

The determination of oxalate in urea is important in the diagnosis of hyperoxaluria

$$\text{oxalate} \xrightarrow{\text{oxalate decarboxylase}} CO_2 + \text{formate}$$

Phosphate and sulphate, which are usually present in urine, inhibit this enzyme.

$$\text{oxalate} \xrightarrow{\text{oxalate oxidase}} 2CO_2 + H_2O_2$$

This enzyme is also inhibited by some anions.
    For urea, see Section 8.4.1.3.

8.4.4 Iodine-selective

For glucose, see Section 8.3.1.

$$\text{L-phenylalanine} \xrightarrow{\text{LAAO PO}} H_2O_2$$

$$H_2O_2 + 2I^- + 2H^+ \longrightarrow I_2 + 2H_2O$$

L-Aminooxidase and peroxidase are co-immobilised in a polyacrylamide gel on the surface of an iodide electrode. However, this sensor suffers more interference and selectivity problems than the ammonia-based sensor in Section 8.4.2.

8.4.5 Ag^+/S^{-2}-linked

$$Ag^+ + xR\text{—}S^{y-} \longrightarrow Ag(R\text{—}S)_x^{(1-xy)-}$$

The above reaction provides a direct potentiometric non-enzymatic method of analysis for cysteine, but it is not totally selective.

$$\text{cysteine} \longrightarrow \text{cystine (non-enzymatic)}$$

The reaction

$$\text{cysteine} + CN^- \xrightarrow{\beta\text{-cyanoalanine synthesase}} HS^- + \beta\text{-cyanoalanine}$$

is more specific, but the cyanide interferes with the electrode.

8.5 REFERENCES

W. J. Albery and D. H. Cranston (1987) 'Amperometric enzyme electrodes: theory and experiment'. in A. P. F. Turner, I. Karube and G. S. Wilson (Eds), *Biosensors: Fundamentals and Applications*, Oxford University Press, Oxford, Chap. 12, pp. 180–210.

I. M. Banat, A. Marchant, P. Nigam, S. J. S. Gaston, B. A. Kelly and R. Marchant (1994) 'Production, partial characterisation and potential diagnostic use of salicylate hydroxylase', in press.

P. N. Bartlett (1987) 'The use of electrochemical methods in the study of modified electrodes', in A. P. F. Turner, I. Karube and G. S. Wilson (Eds), *Biosensors: Fundamentals and Applications*, Oxford University Press, Oxford, Chap. 13, pp. 211–246.

M. F. Cardosi and A. P. F. Turner (1987) 'The realisation of electron transfer from biological molecules to electrodes', in A. P. F. Turner, I. Karube and G. S. Wilson (Eds), *Biosensors: Fundamentals and Applications*, Oxford University Press, Oxford, Chap. 15, pp. 257–275.

A. E. G. Cass, G. Davis, G. D. Francis, H. A. O. Hill, W. J. Aston, I. J. Higgins, E. V. Plotkin, L. D. Scott and A. P. F. Turner (1984) 'Ferrocene mediated enzyme electrode for amperometric determination of glucose', *Anal. Chem.* **56,** 657.

J. F. Cassidy, C. Clinton, W. Breen, R. Foster and F. O'Donohue (1993) 'Novel electrochemical device for the determination of cholesterol or glucose', *Analyst*, **118,** 415.

L. C. Clark, Jr (1987) 'The enzyme electrode', in A. P. F. Turner, I. Karube and G. S. Wilson (Eds), *Biosensors: Fundamentals and Applications*, Oxford University Press, Oxford, Chap. 1, pp. 3–12.

G. Davis (1987) 'Cyclic voltammetry studies of enzymatic reactions for developing mediated biosensors', in A. P. E. Turner, I. Karube and G. S. Wilson (Eds), *Biosensors: Fundamentals and Applications*, Oxford University Press, Oxford, Chap. 14, pp. 247–256.

S.-O. Enfors (1987) 'Compensated enzyme-electrode for *in situ* process control', in A. P. F. Turner, I. Karube and G. S. Wilson (Eds), *Biosensors: Fundamentals and Applications*, Oxford University Press, Oxford, Chap. 19, pp. 347–355.

E. A. H. Hall (1990) *Biosensors*, Open University Press, Milton Keynes.

P. I. Hilditch and M. J. Green (1991) 'Disposable electrochemical biosensors', *Analyst*, **116,** 1217.

J. Jones, R. Marchant and Zhou (1994) Unpublished results, University of Ulster.

S. S. Kuan and G. G. Guilbaut (1987) 'Ion-selective electrodes and biosensors on ISEs', in A. P. E. Turner, I. Karube and G. S. Wilson (Eds), *Biosensors: Fundamentals and Applications*, Oxford University Press, Oxford, Chap. 9, pp. 135–152.

J. F. Rusling and K. Ito (1991) 'Voltammetric determination of electron transfer rate between an enzyme and a mediator', *Anal. Chim. Acta*, **252,** 23.

F. W. Scheller, D. Pfeifer, F. Schubert, R. Renneberg and D. Kirsten (1987) 'Application of enzyme-based amperometric biosensors to the analysis of "real" samples', in A. P. E. Turner, I. Karube and G. S. Wilson (Eds), *Biosensors: Fundamentals and Applications*, Oxford University Press, Oxford, Chap. 18, pp. 315–346.

A. P. F. Turner, W. J. Aston, I. J. Higgins, J. M. Bell, J. Colby, G. Davis and H. A. O. Hill (1984) 'Carbon monoxide: acceptor oxidoreductase from *Pseudomonas thermocarboxydovorans* strain C2 and its use in a carbon monoxide sensor', *Anal. Chim. Acta*, **163,** 161.

R. Wilson and A. P. F. Turner (1992) 'Glucose oxidase: an ideal enzyme', *Biosensors Bioelectron.*, **7,** 165.

L. Ye, M. Hammerle, A. J. J. Olstehoorn, W. Schumann, H.-L. Schmidt, J. A. Duine and A. Heller (1993) 'High density "wired" quinoprotein glucose dehydrogenase electrode', *Anal. Chem.*, **65,** 238.

Chapter 9

Experimental Examples

9.1 INTRODUCTION

The experiments with biosensors described in this chapter have all been used with classes of final honours year undergraduates on the BSc Applied Biochemical Sciences Course at the University of Ulster. They would also be suitable for taught postgraduate courses such as Masters courses. They have each been designed to be completed in one three-hour laboratory period, but could with advantage be staged over more than one period, thus permitting preparation of the biosensors including immobilisation procedures.

The aim is to allow students to experience several aspects of biosensors and their applications. There is one based on glucose oxidase illustrating the principle of mediation, another illustrates an amperometric transducer and another a potentiometric transducer. It is hoped in due course to have one illustrating an optical transducer. The construction, the analytical use and the calibration of biosensors are illustrated.

9.2 GLUCOSE BIOSENSOR (VOLTAMMETRIC)

The first experiment inolved glucose oxidase coupled to ferrocene, but not in a strictly biosensor arrangement. A solution is set up containing a phosphate buffer, glucose and ferrocene. A cyclic voltammogram is obtained using a glassy carbon electrode which shows the reversible cyclic voltammogram typical of ferrocene. Then a measured amount of glucose oxidase is added, the cyclic voltammetry is repeated and the catalytic curve is seen, which is characteristic of the glucose–glucose oxidase–ferrocene system.

This experiment gives experience of cyclic voltammetry and also of the catalytic effect of an enzyme. It is based on a paper by Rusling and Ito (1991) in which peak current data are matched with a computer model to give catalytic rate constants. We did not go that far, but did extract data on peak current ratios which were plotted against the reciprocal of the square root of the sweep rate, which gave a trend similar to that obtained by Rusling and Ito. The following is an account of our experimental procedure and some typical data obtained by the students.

Voltammetric study of the interaction between an enzyme and a mediator

Reference
J. F. Rusling and K. Ito (1991) *Anal. Chim. Acta*, **252**, 23.

Background
Enzymes do not exchange electrons readily with electrodes. However, the addition of an electron mediator can greatly facilitate the process. An electron transport mediator can be used instead of oxygen. Thus, with glucose and glucose oxidase:

$$\text{glucose} + \text{GOD}_{Ox} \longrightarrow \text{gluconolactone} + \text{GOD}_{Red}$$

$$\text{GOD}_{Red} + 2\,\text{Fe(Cp)}_2\text{R}^+ \longrightarrow \text{GOD}_{Ox} + 2\text{Fe(Cp)}_2\text{R} + 2\text{H}^+$$

$$2\,\text{Fe(Cp)}_2\text{R} \longrightarrow 2\,\text{Fe(Cp)}_2^+ + 2\text{e}^- \ (\text{at electrode})$$

We start by studying the cyclic voltammetry of the ferrocene [Fe(Cp)$_2$R] species and then study the effect of adding GOD and glucose to the solution.

Equipment and materials
Bioanalytical Systems CV1A cyclic voltammeter, glassy carbon electrode (soaked overnight in 50 mM glucose), SCE reference electrode, Pt counter electrode, *X–Y* recorder, digital voltmeter, cell.

Phosphate buffer (pH 7.0), 1.0 M glucose, 5 mM ferrocene carboxylic acid, glucose oxidase (GOD).

Procedure
1. Measure the activity of the glucose oxidase by spectrophotometry at 450 nm. Take the molar absorptivity, ϵ, of the absorbing factor (FAD) to be 1.31×10^4 dm^3 mol^{-1}. Dissolve (say) 10 mg of GOD in 5 cm^3 water and measure the absorbance (A) at 450 nm. Then use $A = \epsilon C$ (for a 1 cm cell).
2. Prepare a solution containing 50 mM glucose and 0.5 mM ferrocene carboxylic acid in the pH 7.0 buffer: take 1.0 M glucose (1.25 cm^3) + 5 mM ferrocene carboxylic acid (2.5 cm^3) and dilute to 25 cm^3 with pH 7.0 buffer.
3. Place the three electrodes in the cell through the rubber bung and add the above solution. Deoxygenate the solution with nitrogen for 15 min.

4. Record the cyclic voltammogram of this solution between 0 and + 0.65 V vs SCE at a sweep rate of $20\,mV\,s^{-1}$. Use a current setting of $50 \times \mu A\,V^{-1}$. Measure the following data from the cyclic voltammogram: $E_{p(c)}$; $E_{p(a)}$; $i_{p(c)}$; $i_{p(a)}$. Repeat for several sweep rates: 10, 20, 30, 40 and $50\,mV\,s^{-1}$.
5. Add sufficient GOD to the solution to make its concentration about 10 μM (as determined initially). Weigh out the solid GOD.
6. Record the cyclic voltammogram again. Measure the wave height (i_{cat}) and the half-wave potential ($E_{1/2}$).
7. Repeat the CV at exactly the same scan rates as in (4) and again measure i_{cat} and $E_{1/2}$.

Results and discussion
1. Work out $i_{p(c)}/i_{p(a)}$ values from the CV data from (4) and comment on their values.
2. What is the effect of changing the sweep rate?
3. Why does the wave shape change on adding the enzyme?
4. Work out values of $i_{cat}/i_{p(a)}$ for each sweep rate and plot a graph of these values against $1/(\text{sweep rate})^{1/2}$ ($1/v^{1/2}$). Comment on this graph.

Figure 9.1 shows cyclic voltammograms for this in solution experiment. Some typical data are given in Table 9.1. A graph of $i_{cat}/i_{p(a)}$ versus $1/v^{1/2}$ is shown in Figure 9.2.

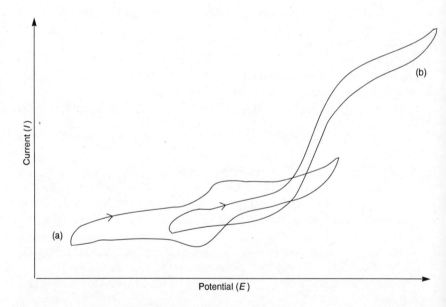

Figure 9.1   (a) Cyclic voltammogram of ferrocene monocarboxylic acid with glucose: (b) the same with added glucose oxidase

Table 9.1 Peak current data at different sweep rates from cyclic voltammograms of ferrocene–glucose with and without added glucose oxidase

$v/V\ s^{-1}$	$1/v^{1/2}$	$i_{cat}/\mu A$	$i_{p(a)}/\mu A$	$i_{cat}/i_{p(a)}$
0.02	7.07	1.55	0.35	4.43
0.03	5.77	1.50	0.475	3.16
0.04	5.00	1.50	0.638	2.35
0.05	4.47	1.80	0.975	1.85
0.06	4.10	1.475	0.975	1.51

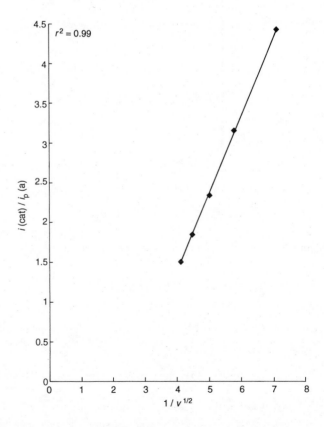

Figure 9.2   Glucose oxidase catalysis of glucose oxidation with ferrocene

9.3 UREASE BIOSENSOR (POTENTIOMETRIC)

The next experiment was to make a potentiometric biosensor. The easiest (and the first to be made) is the urease biosensor (Figure 9.3). The simplest ISE is the glass pH electrode. Unfortunately, this does not give the best urea–urease biosensor, but it is so much more readily available in the laboratory. A pH electrode will work if one uses a histidine buffer. However, it is not easy to immobilise the urease on the glass electrode. In fact, one can demonstrate the effect with urease just added to the solution, as with the first glucose oxidase–ferrocene experiment. The response is fairly small (7 mV per decade), but it is linear in terms of millivolts per decade of urea concentration. Eventually satisfactory immobilization was obtained with gelatine and glutaraldehyde. A membrane can be added to prolong the lifetime.

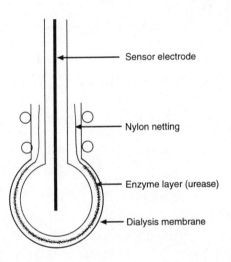

Figure 9.3   Diagram of a urease sensor based on a pH electrode

A UREASE ELECTRODE BASED ON A PH ELECTRODE

References
1. R. Tor and A. Freeman (1986) *Anal. Chem.*, **58,** 1042.
2. P. Durand, A. David and D. Thomas (1978) *Biochim. Biophys. Acta*, **527,** 277.

Background
Urea is easily hydrolysed in the presence of the enzyme urease:

$$(NH_2)_2CO + H_2O \xrightarrow{\text{urease}} 2NH_3 + CO_2$$

The reaction may be followed by using an ammonia-selective electrode, an ammonium ion-selective electrode or a carbon dioxide electrode. With a suitable buffer it may also be followed using an ordinary glass pH electrode. In this experiment, the urease is immobilised on the surface of a glass pH electrode with gelatine cross-linked with glutaraldehyde.

Equipment and materials
Prepared urease electrode based on pH electrode, 0.1 M histidine buffer, 10^{-3} M urea in histidine buffer, SCE electrode, digital voltmeter and an unknown solution of urea.

Preparation of urease/pH electrode
1. Make a gelatine solution (such as 240 bloom) taking 0.5 g in 10 cm³ of solution at 50°C.
2. Add 2.5 mg of urease at 25°C.
3. Pour the mixture on to a glass electrode rotating at ca 100 rpm.
4. Dry at 25°C for 4 h.
5. Immerse in 2.5% glutaraldehyde in 0.02 M phosphate buffer (pH 7.4) for 15 min.
6. Rinse in buffer for 1 h.
7. Store in 0.1 M histidine buffer for at least 4 h.
8. (Optional) cover with a dialysis membrane.

Procedure
1. Incubate the electrode in histidine buffer (5 cm³) in a thermostated cell at 30°C. Measure the potential against the SCE reference electrode at intervals until it is steady.
2. Make a series of solutions of urea in histidine from the stock solution in the range 1, 2, 5, 10, 15 and 20 × 10^{-5} M.
3. Thermostat the cell at 30°C and measure the potential of each.
4. Repeat with the unknown.

Results and discussion
1. Plot two graphs of response (mV) against (i) concentration of urea and (ii) log[urea].
2. Which one is linear? Measure and record the slope.
3. From the graph, determine the unknown urea concentration.
4. Comment on the response time (time for the potential to stabilise).
5. What is the recovery time?
6. Discuss the performance factors of this electrode compared with literature data for urea electrodes.

Some typical data are given in Table 9.2 and the calibration graph is shown in Figure 9.4.

Table 9.2 Some typical potential–concentration data from a urease biosensor based on a pH electrode.

Urea concentration/10^{-5} M	Log[urea]	E/mV
2.5	−4.6	86.06
5.0	−4.3	83.00
10.0	−4.0	77.38
15.0	−3.8	74.46
50.00	−3.3	62.04
Unknown (32.0)	(−3.4)	65.03

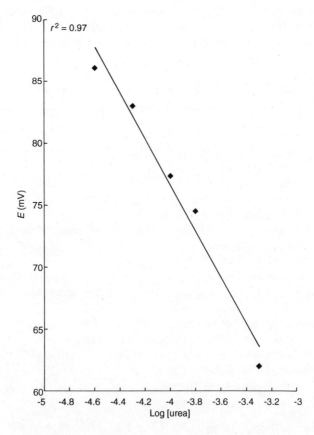

Figure 9.4 Calibration graph for a pH–urease sensor

Slope = 18.9 mV per decade.
Concentration of unknown = 3.2 × 10⁻⁴ M.
Response time = 30–40 s.
Recovery time = 10–15 s.

9.4 TISSUE-BASED BIOSENSOR (VOLTAMMETRIC)

The easiest biosensor to construct was based on vegetable tissue from banana—the 'bananatrode'. This was originally developed for the detection of dopamine, a chemical in the brain. In fact, the polyphenolase enzyme complex in banana is effective with any catechol-type compound, i.e. a 1,2-dihydroxybenzene. The polyphenolases catalyse the oxidation of the catechols, by ambient oxygen, to *o*-quinones. The quinones can be readily detected by electrochemical reduction at a carbon paste electrode. One can combine the enzymatic stage with the electrochemical stage in a banana biosensor by mixing the banana with the carbon paste. The biosensor can be calibrated with simple catechol. An interesting application is the determination of catechol-type substances found in beers. This has been a most successful biosensor for student use. We have used differential pulse voltammetry and the multiple standard addition method of calibration. We have also developed an amperometric method needing much simpler equipment, which could be set up in most school laboratories.

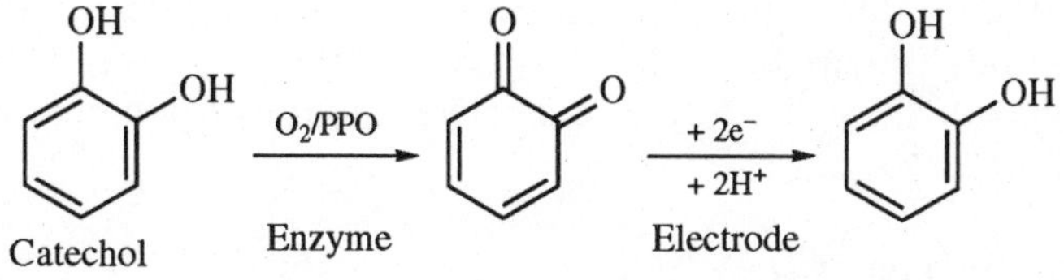

Figure 9.5   Reaction of catechol in a biosensor

The following is based on the students' handout sheet for this experiment.

The 'bananatrode'. Determination of catechols in beer with a plant tissue–carbon paste electrode biosensor using a multiple standard addition method

References
1. J. S. Sidwell and G. A. Rechnitz (1985) *Biotechnol. Lett.*, **7**, 419.
2. J. Wang and M. S. Lin (1988) *Anal. Chem.*, **60**, 1545.
3. D. Madigan, I. McMurrough and M. R. Smyth (1994) *Analyst*, **119**, 863.

Background

Animal or plant tissues and microbial cells can often be used in place of isolated enzymes to catalyse biochemical reactions. Biosensors constructed from such materials usually have a longer lifetime than those made from isolated enzymes and retain a higher enzyme activity.

Dopamine is an important substance connected with cerebral activity and changes in its concentration have been measured *in vivo* in rats via implanted electrodes.

Dopamine is a catechol derivative (Figure 9.6) and can be electrochemically oxidised to dopamine quinone and the quinone reduced back again. It can also be oxidised by oxygen catalysed by the enzyme

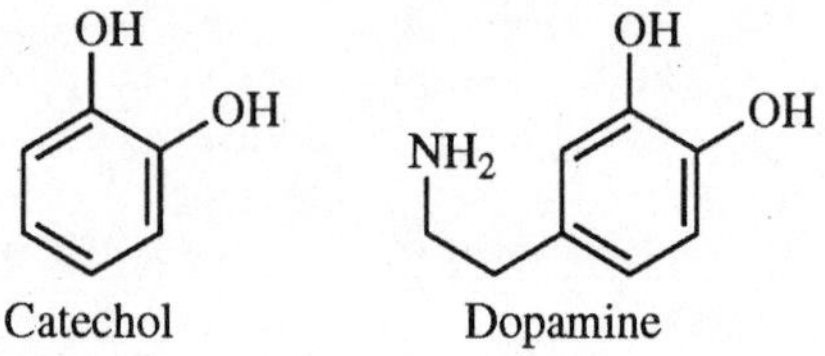

Catechol   Dopamine

Figure 9.6   Structures of catechol and dopamine

polyphenol oxidase, which is contained in bananas (and also in aubergines, apples and potatoes). Sidwell and Rechnitz (1985) developed a banana-based biosensor using a Clark-type oxygen electrode to measure the oxygen uptake. Wang and Lin (1988) simplified this by making a carbon powder–banana paste electrode (Figure 9.7). The electrode is used to measure the reduction of the quinone. They discovered that catechol and hydroquinone responded equally well.

Equipment and materials

Bruker pulse polarograph, *X–Y* recorder, digital voltmeter, electrode for carbon paste, SCE, Pt electrode, Acheson grade 38 graphite powder, mineral oil (Nujol, liquid paraffin), fresh banana, catechol, phosphate buffer (pH 7.4), sample of beer (e.g. Miller).

Procedure

1. Prepare a carbon–banana paste electrode. Cut a 5 mm thick section of banana pulp from the centre of the fruit with a spatula. Crush a 0.11 g portion with a spatula and mix it with 0.9 g of mineral oil. Add 1.1 g of graphite powder (Acheson grade 38) and mix thoroughly. Place a portion of this paste in the end of an electrode holder. Smooth the surface of the electrode paste on a smooth card before each run.
2. Prepare standard solutions of (a) catechol (0.01 M) and (b) 0.05 M

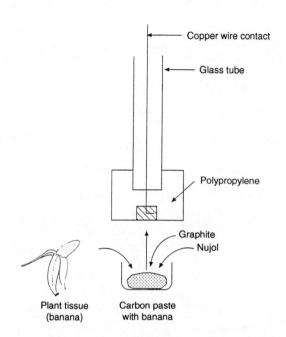

Figure 9.7   Banana-based electrode—'bananatrode'

phosphate buffer (pH 7.4) using deionised water.
3. Set up the Bruker pulse polarograph in the differential pulse mode to sweep between +0.2 and −0.4 V at a 10 mV s⁻¹ scan rate with a pulse amplitude of 25 mV. Try a current sensitivity of 5 μA full-scale.
4. Run a differential pulse polarogram of the buffer alone (25 cm³), sweeping between +0.2 and −0.4 V.
5. Add an equal volume (25 cm³) of the beer sample and repeat the scan.
6. Make five successive additions of 0.1 cm³ aliquots of a standard solution of catechol (5 mM) and repeat the scan after each addition.

Results and discussion
1. Plot a multiple standard addition calibration graph of peak current (i_p), measured above the blank line, against added concentration (Δc).
2. Measure the slope and determine the correlation coefficient of the linear part of the graph.
3. Measure the negative intercept at $i_p = 0$ and hence estimate the catechol content of the beer.
4. Comment on the performance criteria of the bananatrode.

5. What is present in beer to give a catechol reaction? (see Madijan *et al.*, 1994).

Some typical results are presented. Figure 9.8 shows some differential pulse polarograms, Table 9.3 gives the data and Figure 9.9 shows the calibration graph.

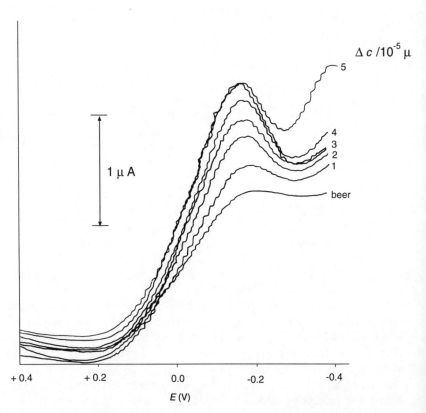

Figure 9.8 Differential pulse polarograms of beer and added concentrations of catechols

The calibration slope is 28.3 A mM^{-1} with a correlation coefficient of 0.992; the negative intercept of 5.11 × 10^{-5} gives a catechol concentration in the Miller beer of (1.02 ± 1.0) × 10^{-4} M (allowing for the initial 1+1 dilution).

The response time is controlled by the sweep rate to 60 s. A faster response can be obtained with an amperometric method. The electrode is stable and gives reproducible data for at least 2 weeks.

Table 9.3 Data for peak current vs concentration for the determination of catechols in Miller beer

Catechol standards/10^{-5} M	i_p/A
0.00 (beer)	1.45
0.998	1.8
1.99	2.0
2.98	2.3
3.95	2.7
4.93	2.8

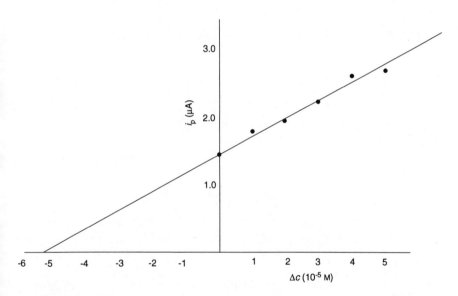

Figure 9.9　Multiple standard addition calibration graph for catechol in beer

9.5　REFERENCES

P. Durand, A. David and D. Thomas (1978) 'An enzyme electrode for acetylocholine', *Biochim. Biophys. Acta*, **527**, 277.

E. A. H. Hall (1990) *Biosensors*, Open University Press, Milton Keynes.

D. Madigan, I. McMurrough and M. R. Smyth (1994) 'Determination of proanthocyanidins and catechins in beer and barley by high-performance liquid chromatography with dual-electrode electrochemical detection' *Analyst*, **119**, 863.

J. F. Rusling and K. Ito (1991) 'Voltammetric determination of electron transfer rate constant between an enzyme and a mediator', *Anal. Chim. Acta*, **252**, 23.

J. S. Sidwell and G. A. Rechnitz (1985) '"Bananatrode"—an electrochemical biosensor for dopamine', *Biotechnol. Lett.*, **7**, 419.

R. Tor and A. Freeman (1986) 'New enzyme membrane for enzyme electrodes', *Anal. Chem.*, **58**, 1042.

J. Wang and M. S. Lin (1988) 'Mixed plant tissue–carbon paste electrode', *Anal. Chem.*, **60**, 1545.

Chapter 10

Commercial Applications

10.1 MEDICAL APPLICATIONS

The major advances in sensor technology have come through the needs of medical care. The analysis of body fluids for a range of components is essential in modern medicine. Hitherto most of these analyses have been done by taking samples of blood, urine, etc., and submitting them to medical laboratory scientists for conventional wet chemical analysis. Usually the results took several days to obtain, check and report back to the doctors for assessment. This may be satisfactory in some situations, but is completely useless for others, such as with patients in intensive care, whose condition can change from minute to minute. With diabetics also, rapid analyses of blood glucose are needed.

Sensors have been developed to provide portable, rapid, on-the-spot analyses, in some cases giving continuous on-line monitoring. We can consider a number of different clinical situations:

(i)   Intensive care units.
(ii) Casualty/emergency rooms, general wards.
(iii) GP consulting rooms.
(iv) Home use by the patient.

The number of different assays required regularly is fairly limited. In blood these are oxygen, carbon dioxide, pH, lactate, amino acids and ketones; in blood or sub-cutaneous tissue, glucose, sodium, potassium, calcium, creatinine and urea; and in urine, sodium, potasium, creatinine and urea.

In addition to these regular assays, in emergencies a number of other instant assays are often required. These include amylase, paracetamol, salicylate, creatine kinase, aspartate aminotransferase and ammonia, in addition to some already mentioned, i.e. glucose, sodium, potassium, calcium, creatinine, urea, oxygen, carbon dioxide and pH.

The commonest need in emergencies is to check on and distinguish between paracetamol and salicylate (aspirin) overdoses, as the former can cause irreversible liver damage at relative low overdoses. Sometimes other drug overdoses need to be determined, such as phenobarbitone or paraquat. There are other very important assays which would become much easier with a biosensor if one could be developed, e.g. for hepatitis B antigen, the related HIV (AIDS) virus antigen and the heart active digoxin.

In the doctor's surgery, it is becoming increasingly useful to be able to obtain instant diagnostic information rather than waiting for laboratory tests. Tests for glucose are now often available. Others which would be very useful are amylase, creatine kinase, cholesterol and triglycerides in blood, and also creatine, potassium and the viral antigens.

The major sensor used by a patient at home is the glucose biosensor for diabetics. The Exactech instrument marketed by Medisense is very compact (in fact, the first pen-like version proved to be too compact, in that some diabetics suffer from impaired vision) and inexpensive—the measurement unit currently (1995) markets at £45 ($70) and the disposable electrode/sensor strips cost 40p ($0.60).

10.1.1 'Artificial Pancreas'

This is perhaps the most glamorous and sophisticated application of biosensors, and is the dynamism for much of the very considerable amount of research into glucose biosensors. People with diabetes have a number of requirements to enable their metabolisms to run in a stable manner. Overall the requirement is to maintain a balance between carbohydrate intake and insulin production by the pancreas. Insulin is a hormone essential for the metabolism of carbon sources and is not produced in sufficient amounts in diabetics. This results in an abnormally high level of glucose in the blood, which must be measured regularly. The conventional method for blood glucose determination is for samples of blood to be removed periodically and taken to a laboratory for analysis for glucose (and sometimes analysis for various ketonic species). Then injections of insulin are made periodically, to compensate for the inability of the pancreas itself to produce sufficient insulin. In addition, the patient has to control her/his diet to conform with the body's carbon intake requirements. The diet may be low in glucose, although occasionally extra amounts of sugar-containing materials need to be taken.

The problems are the inconvenience for the patient in carrying out these requirements and that the periodicity of the injections causes an undesirable periodicity in the levels of insulin and consequently of glucose, which results in a potentially unstable metabolism. Further, the doses of insulin are generally related to a regular prescription and not to the immediate requirements of the body's metabolism.

The first requirement to deal with these problems is a nearly continuous or, better, on-line measurement of blood glucose. This requires a sensor implanted subcutaneously which can be left in place for at least days, and preferably weeks at a time.

The second requirement is to convert the glucose level into an insulin requirement level.

The third requirement is continuous injection of insulin, via an insulin pump, at a rate determined by the glucose level. These last two requirements must be linked together by a fool-proof, fail-safe microprocessor device so that the amount of glucose determined minute by minute (or continuously) controls the rate of absorption of insulin.

A biosensor which will fit these demands must have a number of basic properties:

(i)   It must give a linear response in the range 0–20 mM ± 1 mM.
(ii) It must be specific for glucose, and unaffected by changes in the concentrations of other metabolites and slight variations in ambient conditions.
(iii) It must be biocompatible.
(iv) It must be small so as to cause minimal tissue damage during insertion.
(v)   It must be capable of external calibraiton to within <10% in 24 h.
(vi) The response time must be less than 10 min.
(vii) It must have a lifetime of several weeks in use.

The three basic schemes for insulin therapy, shown in Figure 10.1, are as follows.

(a) *Conventional:* multiple insulin injections related to manual glucose measurement with colour-reactive strips.
(b) *Open loop:* continuous preprogrammed insulin infusion with an insulin pump dependent on manual glucose measurement using a biosensor.
(c) *Close loop:* continuous self-regulated insulin infusion regulated by the continuously monitored glucose level using a biosensor, the two linked by a microprocessor.
(d) *Additionally, telephone modem:* instead of a continuous controlled loop operation, the patient would be linked up to a microprocessor via a telephone modem to a remote microprocessor at a laboratory or surgery.

10.2 BIOTECHNOLOGY—FERMENTATION PROCESS CONTROL

The biotechnology industry is a greatly expanding one which requires considerable analytical monitoring, some of which could potentially be

188

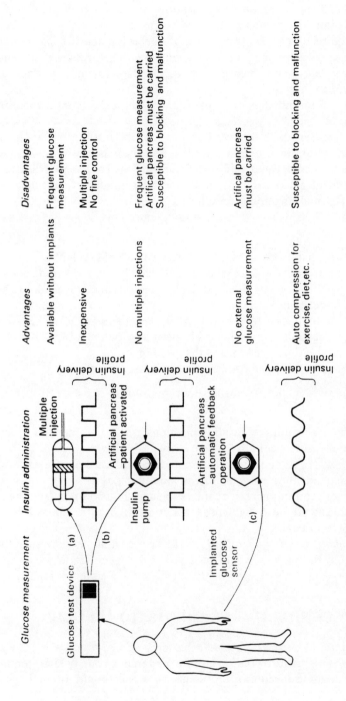

Figure 10.1 Schemes for insulin therapy. Reproduced by permission of the Open University Press from Hall (1990)

carried out using biosensors. There are two main areas of need: one is concerned with monitoring the active components and products of a biotechnology (usually fermentation) process, and the other is concerned with analysing for pollutants and microbial contaminants. Biosensors are highly selective for enzymes and immunological components. They have the ability to target particular molecules in a highly complex soup of other molecules.

Biosensors do present certain problems, however. They are very sensitive and are susceptible to changes in temperature, pH and osmotic pressure. Further, they are subject to fouling and they deteriorate with time. For example, food processing often involves a steam sterilisation process, which would destroy the biological component of a biosensor.

One can consider three time zones for the application of analytical sensors, as demonstrated in Figure 10.2.

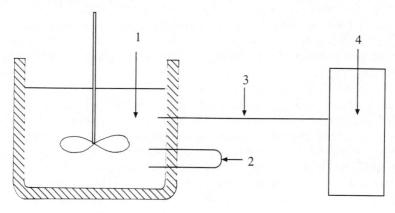

1. On-line, real time
2. On-line with feedback loop
3. Off-line, local
4. Off-line, distant (laboratory)

Figure 10.2   Time zones for the application of analytical sensors

The most immediate and demanding is on-line, real-time monitoring with feedback control. So far not many variables have been measured at this level. The obvious ones are temperature, pH, CO_2 and O_2, none of which normally involves biosensors. If however, one could measure carbon sources, products, dissolved gases, etc., in the real-time mode there could be considerable optimisation of the procedure, resulting in increased yields of products at decreased material cost. There would be improved product quality, and so fewer rejections. Quality variation in raw materials could be tolerated more easily, as the variation could be compensated for via the process control feedback loop. There would also be less need for human judgement of the progress of the fermentation by 'the seat of the pants'. The energy needs could be optimised. The overall performance of the plant would be improved and dead time would be eliminated.

The second time zone is off-line local. This can result in fine control with a short time lapse. There is probably the most scope for biosensors in this area at present. It probably involves sampling the batch and making on-the-spot measurments. From these results, manual adjustment to the operating parameters can be made.

The third time zone is off-line distant. This usually means again taking samples, and sending them to a central laboratory. This results in coarse control with a significant lapse of time. This mode would permit much more sophisticated and elaborate analytical techniques such as HPLC and GC–MS. Such methods would normally need to be available as a check on the process.

The best known industries using biotechnology are the food, drink and pharmaceutical industries. However, such has been the expansion of this area that there are many other less likely sounding applications. However, to illustrate some examples of actual and potential applications of biosensors, we shall consider the food and drink industries.

The food and drink industries are continually in need of analytical methods both for compounds that have not previously been monitored and to replace inefficient and/or expensive procedures. Biosensors need to be inexpensive, reliable and robust under difficult conditions. Also needed are speed of operation, short sample preparation times, less need for skilled labour and replacement of unacceptable organic reagents. Analytical procedures are needed for the detection of contaminants, checking product contents, monitoring raw material conversion and evaluation of product freshness.

Governments and consumers require increasing levels of product scrutiny for contaminants, particularly toxins, microorganisms such as *Salmonella* and *Listeria* and pesticides in many products such as milk, wine and fruit juice.

Another common requirement is that the components of a food product should be listed—particularly vitamins, essential amino acids, colouring agents, emulsifiers, flavouring agents, preservatives, antimicrobial agents and allergens.

Freshness of products, especially fish and meat, is now of great concern, especially raw fish, which is must eaten in oriental regions. Indicators of freshness include alcohol in fruit, aldehydes in fat, histamine in fish, lactate in canned vegetables and lactose in milk.

10.2.1   Biosensors Useful in Food and Drink Analysis

Biosensors potentially useful in such analyses are listed in Table 10.1.

Immunological assays are needed for some analyses. They can usually be

Table 10.1 Amperometric biosensors that could be relevant to the food and drink industries [reproduced by permission of Elsevier from Luong *et al.* (1991)]

Type of sensor	Target compound	Biological material
Hydrogen peroxide based	Sulphite	Sulphite oxidase
	Xanthine	Xanthine oxidase
	Cholesterol	Cholesterol oxidase
	Glutamate	Glutamate oxidase
	Alcohol	Alcohol oxidases
	L-Amino acids	L-Amino acid oxidases
	Sucrose	Invertase, mutarotase, glucose oxidase
	Inosine phosphate	Nucleoside phosphorylase, xanthine oxidase
	Aspartame	Peptidase, aspartate aminotransferase, glutamate oxidase
	Glutamine	Glutaminase, glutamate oxidase
	Lactate	Lactate oxidase
Oxygen based	Glucose	*Pseudomonas fluorescens*
	Acetic acid	*Trichosporon brassicae*
	Ethanol	*Trichosporon brassicae*
	Hydrogen peroxide	Bovine liver
	Tyrosine	Sugar beet
	Glutamic acid	Glutamate oxidase
	Lysine	*Trichoderma viride*
	Sulphite	Sulphite oxidase
	Oxalate	Oxalate oxidase
	L-Amino acids	L-Amino acid oxidases
	Monoamine	Monoamine oxidase

enzyme linked and may be used with amperometric, optical, thermometric or piezoelectric transducers, especially for bacterial contamination.

10.2.2 Commercially Available Biosensors

The market in biosensors is a constantly changing one. The first commercial biosensor was marketed in 1974 by the Yellow Springs Instrument Company (YSI). It was a glucose biosensor based on the Clark oxygen electrode. Similar devices have been developed in Japan, Germany and Switzerland. Subsequently, YSI appear to have withdrawn from this market, perhaps owing to the more recent development of a glucose biosensor based on a ferrocene mediator by Medisense. Most of the earlier commercial biosensors were based on either oxygen reduction using a platinum cathode at -0.6 V with a silver anode or hydrogen peroxide oxidation using a silver anode at +0.6 V with a platinum cathode.

The YSI Model 2700 Industrial Analyser (based on the oxidation of hydrogen peroxide) was used in the sugar, molasses and confectionary industries as a standard method for determining levels of glucose and sucrose. With the same basic transducer, YSI could also supply enzymes for the detection of total starch, fructose, dextrose, lactose, ethanol, glycerol and L-lactate. Other companies offering similar biosensors for glucose include Fuji Electrice (Japan), Kyoto Daichi Karaku (Japan), Omron Toyoba (Japan), Solea-Tacussel (France) and VEB-MLW Prüfgeräte-Medigen (Germany).

A related amperometic biosensor for the determination of fish freshness is made by Pegasus Biotechnology (Canada). Is uses immobilised nucleotidase, nucleoside phosphorylase and xanthine oxidase to convert the degradation products of ATP into uric acid and hydrogen peroxide. Another freshness meter based on oxygen reduction, the KV-101, is made by the Oriental Electric (Japan). A multi-purpose bioanalyser, containing oxidases for determining alcohol, glucose, lactose and lactic acid, was made by Provesta (Oklahoma, USA), but it has now been discontinued. Flow injection analysis (FIA) biosensors have been produced by Control Equipment (New Jersey, USA) and Eppendorf North America (Wisconsin, USA). Some intriguing devices to analyse odours have been developed— 'artificial noses'. Some are not really biosensors but are based on arrays of electrodes with different polymer coatings. However, one marketed by Sogo Pharmaceutical (Japan) uses a synthetic bilayer film on a piezoelectric transducer.

10.3   ENVIRONMENTAL MONITORING

There is great potential for routine and continuous anlayses for impurities and pollutants in water and air. The number of potential analytes is huge. The basic targets are biological oxygen demand (BOD), pH, conductivity, inorganic ions such as nitrate, fluoride, sulphate, phosphate, potassium, sodium and calcium and organic materials such as phenols, urea, pesticides, fertilisers and humic substances.

Three general types of analysis are needed:

  (i)   continuous on-line, real-time monitoring;
  (ii) regular 'alarm' monitoring;
  (iii) occasional random or discrete monitoring.

Many of these analyses are currently done with electrochemical sensors or colorimetric assays.

Biosensors have been used in four ways: (i) enzymes, (ii) microorganisms,

(iii) receptors for photometric methods and (iv) antibodies.

(i) Enzymes

Acetylcholinesterase binds with high affinity to organophosphorus insecticides. Details are discussed in Chapter 6.

(ii) Microorganisms

Biological oxygen demand (BOD) is one of the most important regular assays done on water. It is a meausure of how much oxygen is used up in oxidising all the biological organic matter in the sample. The conventional method requires an incubation period of up to 5 days. However, a biosensor based on the oxygen electrode would be ideal for this assay. The microorganisms *Clostridium butyricum* and *Trichosporon cutaneum* work satisfactorily. They are mounted in front of an oxygen electrode. The electrode is flushed with oxygen-saturated buffer solution and the base oxygen current is read. The sample is then injected and, after stabilisation, the oxygen current is read again. The different in the two readings is linearly proportional to the conventional BOD analysis.

The electrode is designed to work in the temperature range 25–30 °C; however, a BOD sensor has been developed for use at higher temperatures using a thermophilic bacterium isolated from hot springs which is stable at 60 °C and above. This is useful for monitoring hot water outputs from factories.

Mercury(II) can be determined with a luciferase from Japanese pine-cone fish.

(iii) Receptors

α-aminobutyric acid (GABA) binds with cyclodienes, pyrethroids, bicyclophosphates and orthocarboxylates. The muscarinic receptor binds organophosphate insecticides. The aryl hydrocarbon receptor binds dioxins.

(iv) Antibodies

These have been applied to a wide range of analytes, including many herbicides, insecticides, polychlorinated biphenyls (PCBs), dioxins, pentachlorophenol, benzo[a]pyrene, benzene, toluene and xylene.

For environmental monitoring there are several potential advantages in using biosensors. They are compact and therefore portable, for on-the-spot analyses, and they are therefore also relatively inexpensive, even disposable. For example, a biosensor analysis on the spot may cost about £1 compared with over £100 in a laboratory, for which the sample has to be collected, isolated, packaged and transported, involving a chain of several people. The biological component offers high selectivity and usually sufficient accuracy for environmental monitoring. The target performance requirements for biosensors used for environmental analyses have been discussed in Chapter 7 (see Table 7.1), but are repeated here for convenience:

Cost: approx. £0.7–10 per assay
Assay time: 1–60 min
Sampling: Minimal for ground water, soil extracts, blood, urine, saliva, air
Range: >Two orders of magnitude
Sensitivity: ppm–ppb
Specificity: Enzymes/receptors—one or more groups of related compounds
              Antibodies—specific to one compound or one group of
              closely related compounds
Personnel: Minimal training required—1–2 h training period.
Size: Portable by one person, no external power, or carried in a van
              or truck with limited power requirements

These are, of course, broad and general requirements. Specific requirements will depend on the particular sample and application needed.

Ion-selective electrodes, not involving a biological selective material, are commonly used for ions such as H^+ (pH), fluoride, nitrate, calcium and potassium. Phosphate is difficult to determine with an electrode. It is usually assayed colorimetrically, but a biosensor method has been developed as discussed in Chapter 8.

Organic pollutants offer greater potential scope for the use of biosensors. We have already referred to the use of piezoelectric sensors for malathion and parathion in Chapter 6.

Air pollutants are gases such as NO, NO_2, SO_2, O_3 and organic vapours such as octanes and benzene.

10.4   DEFENCE AND SECURITY INDUSTRIES

The military has a strong interest in portable sensors for personal, tank or field-unit use. This sector is possibly the largest potential market for biosensors for the detection and measurement of chemical warfare agents such as nerve gases (e.g. Sarin, used in Japan in 1995) and mustard gas (used in the 1914–18 War). The Thorn-EMI NAIAD nerve gas detector has a multimillion dollar market, as has Dutch Acal from Delft. The development of these sensors is closely related to the political situation. For example, very little use was in fact made of chemical weapons in the 1991 Gulf War. Other applications of more general use would be for the detection of trace vapours, explosives (see Chapter 5 for TNT) and illicit drugs. The acetylcholine receptor systems (see Chapter 2) can be adapted with a matrix of 13–20 proteins to cover about 95% of all toxin detection. Dipstick tests using monoclonal antibodies have been developed by the US Army for the detection of agents such as Q-fever, nerve agents, yellow rain fungus and soman. A photo-immunoassay using luciferase can detect down to 10^{-18} M TNT (see Chapter 5).

10.5 FABRICATION AND MASS PRODUCTION

The main market for biosensors is in diagnostics, which is a very large and expanding market with the increasing concern for preventive medicine. Projected market volumes for clinical testing products in the European market up to the 1990s are shown in Figure 10.3, indicating a market size

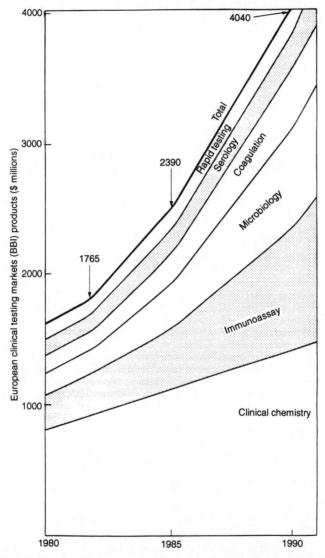

Figure 10.3 The European market for clinical testing products. Source: Biomedical Business International. Reproduced by permission of the Open University Press from Hall (1990)

greater than $4000 million per annum.

The data for biosensors in the USA in particular are shown in Figure 10.4.estimating a figure of over $125 million by 1995, with about $4 million in the UK as shown on the right. The steep rise in the USA could reach $100 million by 2000 A.D. World-wide, there was a demand of about $2000 million ($2 billion) in 1990, rising at up to 45% per annum. About half of these applications are in the clinical area, with the rest divided between agriculture, environment and other industrial process monitoring (such as biotechnology).

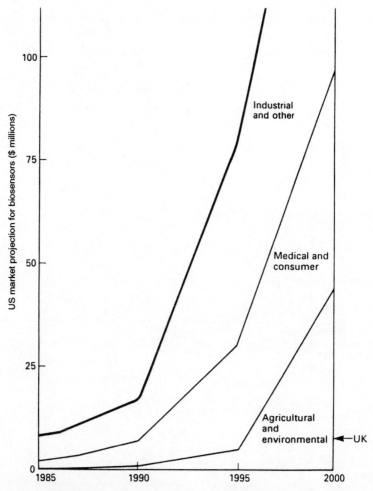

Figure 10.4   US market projection for biosensors, with the UK figure also indicated. Source: Biomedical Business International. Reproduced by permission of the Open University Press from Hall (1990)

Inventing a new biosensor which will work satisfactorily in a laboratory is a long way from manufacturing a product that can be successfully marketed to operate consistently within the specified performance limits, be attractive to the potential user and be marketable at an acceptable price. While the initial costs of inventing the device may be (say) $75 000, the total development costs will probably run into millions of dollars. This section will look at some of the problems to be overcome during this process. Many of them are common to marketing any new product, but we shall try to concentrate on those peculiar to biosensors.

We can recognise a number of stages in the process of the development of a product. They are illustrated in Figure 10.5, known as the product life cycle.

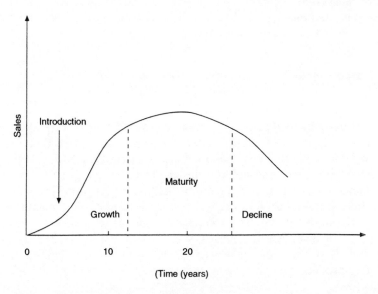

Figure 10.5 Product life cycle for analytical sensors. Reproduced by permission of Oxford University Press from McCann (1987)

The introduction phase is usually carried out by one company and involves laboratory development, prototype testing, regulatory approval from appropriate bodies (especially for medical use) and extensive clinical testing. The company will then need to establish distribution, marketing and servicing networks and advertising to introduce the product to potential consumers. This programme largely determines the success of the product, rather than the intrinsic quality of the product itself. Because of all these processes, the introduction phase for a medical product is usually relatively long. With a completely novel product the company will be in a monopoly position at this stage, helped by patents taken out during the research period.

The growth stage shows a rapid increase in product sales. However, the relatively high initial price soon attracts competitors and competing products which emulate the original product as closely as possible in performance, while avoiding patent infringements. The market for the initial product continues to grow as it becomes more widely known, widely available and develops a history of reliability, and so is widely accepted.

Eventually the market becomes saturated, and sales will slow to the steady region of maturity. At this stage there will probably be a number of competitors selling similar products, perhaps at advantageous prices. The patent protection will become eroded with time and there may be cross-licensing arrangements. There will therefore have to be price cutting and hard-sell marketing.

Eventually the market will begin to decline owing to obsolescence or the introduction of a radical new technology.

10.5.1   Stages in the Development of a New Biosensor

10.5.1.1   FEASIBILITY

At a fairly early stage after the invention of a new product, preferably before going to the expense of taking out patents, preliminary market research needs to be done to see if an adequate market exists to justify development.

A prototype then has to be designed, constructed and tested for performance and particularly reliability. Although the prototype is initially constructed as a one-off, it is necessary to be aware of likely manufacturing costs, particularly if any components (such as enzymes or complex transducers) or manufacturing stages (such as immobilisation procedures) are very costly. If all this looks satisfactory, one can go on to a limited quantity scale-up to produce enough biosensors for testing in the field or clinically.

10.5.1.2   SAFETY CONSIDERATIONS

Safety considerations are of prime importance, especially with biosensors for medical use. Safety during the manufacturing process is also of prime concern, such as during the disposal of large volumes of solvents used in the doping of electrodes. There may be a need to set up expensive clean rooms, not only when using hazadous solvents but also when working with highly pure semiconductor materials.

10.5.1.3   SHELF-LIFE AND PACKAGING

Shelf-like and packaging are of vital importance. Humidity is particularly detrimental to biosensors, especially with strip devices. A shelf-life of at least

18 months is normally required, to permit distribution and any necessary recalibraiton before sale. Medisense use disposable electrode–enzyme strips which are sealed individually into low-humidity foil sachets, which is probably the best system.

10.5.1.4 PRODUCTION

The production process can be carried out in one of three ways, known as job, batch and flow. In job production, a single unit is made by a single operator or a group of operators, who must be capable of performing all tasks. With greater complexity, a batch process is needed in which there is some degree of specialisation. Careful planning is needed. Flow production is like batch production without any rest periods. The rate of material flow is increased and there is a reduced requirement for skills. It can be rather inflexible, in that faults affect the whole scheme and modifications are not easily introduced.

10.5.1.5 COMPONENTS

These need to be readily available, preferably from a variety of sources, and of acceptable quality. For biosensors the key components are the biological materials, such as enzymes, microorganisms and antibodies. Glucose oxidase is a by-product of a fermentation reaction and is widely available commercially from a range of suppliers in large quantities. Aston (1992) attempted to develop a biosensor for methanol in drinking water using methanol dehydrogenase. This was not commercially available and so the group developed their own method of production. However, this dedicated process was not very suitable as a regular supply for the manufacture of a commercial product and then satisfying market demands.

Quality control of material is particularly important with biological materials, as their properties can vary considerably from one source to another and from one time to another. Hence one must determine the maximum tolerances allowable, such as the specific activity of an enzyme, and then check every batch of material before use.

10.6 COMPARISON OF DESIRABLE CHARACTERISTICS IN A MARKETABLE BIOSENSOR

Madou and Tierney (1993) and Griffiths and Hall (1993) have listed the important characteristics in a comparative manner. Madou and Tierney gave two tables showing 'Comparison of biosensors functional characteristics' and

'Comparison of biosensor design'; general comments are given in each table. Griffiths and Hall gave a bar diagram in which each of eight characteristics is given a weighting from 0 to 5 for each of five general types of transducer type. The major categories are sensitivity, cost, selectivity, versatility, range, availability, future and simplicity. In summary, these tables show that the most generally favourable are amperometric biosensors, in which all characteristics except selectivity and versatility are given a score of 5 and the others receive 4. Potentiometric sensors are next most favourable, with all categories rated 4 except versatility and simplicity (5 each). The major drawbacks at present with optical sensors are cost (rated 1) and availability (rated 2). Versatility and future are both rated 5. Piezoelectric and thermal sensors are fairly poorly rated, with most characteristics rated 2 or less.

10.7 REFERENCES

W. J. Aston (1992) 'Manufacturing biosensors', *Biosensors Bioelectron.*, **7**, 161.
W. P. Carey (1994) 'Multivariate sensor arrays as industrial and environmental monitoring systems', *Trends Anal. Chem.*, **13**, 210.
P. R. Coulet (1988) 'Biosensor-based analyzers: from design to instrument', in G. G. Guilbault and M. Mascini (Eds), *Analytical Uses of Immobilised Biological Compounds for Detection, Medical and Industrial Uses*, NATO ASI Series, Reidel, Dordrecht, pp. 319–327.
D. Griffiths and G. Hall (1993) 'Biosensors—what real progress is being made?', *TIBTech*, **11**, 122.
E. A. H. Hall (1990) *Biosensors*, Open University Press, Milton Keynes.
P. D. Home and K. G. M. M. Alberti (1987) 'Biosensors in medicine: the clinician's requirements', in A. P. F. Turner, I. Karube and G. S. Wilson (Eds), *Biosensors: Fundamentals and Applications*, Oxford University Press, Oxford, Chap. 36, pp. 723–736.
J. H. T. Luong, C. A. Groom and K. B. Male (1991) 'The potential role of biosensors in the food and drink industries', *Biosensors Bioelectron.*, **6**, 547.
M. Madou and M. J. Tierney (1993) 'Required technology breakthroughs to assume widely accepted biosensors, *Appl. Biochem. Biotechnol.*, **41**, 109.
E. McCann (1987) 'Exploiting biosensors', in A. P. F. Turner, I. Karube and G. S. Wilson (Eds), *Biosensors: Fundamentals and Applications*, Oxford University Press, Oxford, Chap. 37, pp. 737–746.
V. M. Owen (1988) 'Commerical aspects of the use of immobilised compounds', in G. G. Guilbault and M. Mascini (Eds), *Analytical Uses of Immoblised Biological Compounds for Detection, Medical and Industrial Uses*, NATO ASI Series, Reidel, Dordrecht, pp. 329–339.
K. R. Rogers and J. N. Lin (1992) 'Biosensors for environmental monitoring', *Biosensors Bioelectron.*, **7**, 317.
R. Romito (1993) 'Biosensors: diagnostic workhorses of the future?', *S. Afr. J. Sci.*, **89**, 93.
T. H. Scheper and F. Lammers (1994) 'Fermentation monitoring and process control', *Curr. Opinion Biotechnol.* **5**, 187.

Index

absorbance 1, Ch. 5, 87, 88, 100, 136
acetaldehyde 108, 151, 166
Acetobacter xylinium 164
acetone 151
4-acetylaminophenol (paracetamol) 73, 74, 167, 185, 186
acetylcholine 194
acetylcholinesterase 124
acid
  acetic 21, 151, 165
  D-amino 148
  butyric 21
  formic 21
  glutamic 21
  propionic 21
  ascorbic
    interferant 6, 44
    oxidase 46
    redox potentials 151
    thermister sensor 131
acidity, pH control and measurement 1, 12, 21, 62, 99, 100, 102, 144, 185, 186, 189, 192, 194
  pH electrode (as used in a biosensor) 1, 6, 16, 48, 64, 65, 84, 141, 160, 161, 167, 176
  pH meter 1
Acinetobacter calcoaceticus 27
acridinium-based fluorescent reagent (for halides) 102
action potential 28
adamantyl dioxetine phosphate 105, 106
adenosine 19, 169

deaminase 169
adenosine
  monophosphate (AMP) 19
  triphosphate (ATP) 107, 192
adenosine-5'-phosphosulphate reductase 125
admittance 78
adsorption 125
  see also immobilisation
agriculture 12, 196
AIDS 186
alanine 84, 85, 143
alcohols 12, 191
  see also individual names
alcohol
  dehydrogenase 108, 160
  oxidase 139, 147
aldehyde oxidase 147
alkaline phosphatase 24, 99, 105, 106, 125
allantoin 143
allergens 190
alumina 33, 48
aluminium 62
(3-aminopropyl)triethylsilane 128
amino acids 140, 185, 190, 191
  see also individual names
aminobutyric acid 193
aminooxidase 191
aminotransferase 11
ammonia 6, 11, 16, 22, 46, 85, 100, 101, 140, 169
  see also under ion selective electrodes
  gas sensor 67, 83, 124

DUE DATE

	201-6503		Printed in USA